# Natural Kinds and Conceptual Change

According to the received tradition, the language used to refer to natural kinds in scientific discourse remains stable even as theories about these kinds are refined. In this illuminating book, Joseph LaPorte argues that scientists do not discover that sentences about natural kinds, like 'Whales are mammals, not fish', are true rather than false. Instead, scientists find that these sentences were vague in the language of earlier speakers, and they refine the meanings of the relevant natural-kind terms to make the sentences true. Hence, scientists change the meanings of these terms. This conclusion prompts LaPorte to examine the consequences of this change in meaning for the issue of incommensurability and for the progress of science.

This book will appeal to students and professionals in the philosophy of science, the philosophy of biology, and the philosophy of language.

Joseph LaPorte is an assistant professor of philosophy at Hope College.

*To my wife, Carmelita,*
*and in memory of Bob Jackson*

# Natural Kinds and Conceptual Change

JOSEPH LAPORTE

*Hope College*

PUBLISHED BY THE PRESS SYNDICATE OF THE UNIVERSITY OF CAMBRIDGE
The Pitt Building, Trumpington Street, Cambridge, United Kingdom

CAMBRIDGE UNIVERSITY PRESS
The Edinburgh Building, Cambridge CB2 2RU, UK
40 West 20th Street, New York, NY 10011-4211, USA
477 Williamstown Road, Port Melbourne, VIC 3207, Australia
Ruiz de Alarcón 13, 28014 Madrid, Spain
Dock House, The Waterfront, Cape Town 8001, South Africa

http://www.cambridge.org

First published 2004

Printed in the United States of America

*Typeface* Times Roman 10.25/13 pt.     *System* LaTeX $2_\varepsilon$   [TB]

*A catalog record for this book is available from the British Library.*

*Library of Congress Cataloging in Publication Data*

LaPorte, Joseph.
   Natural kinds and conceptual change / Joseph LaPorte.
      p.   cm. – (Cambridge studies in philosophy and biology)
   Includes bibliographical references (p. ) and index.
   ISBN 0-521-82599-7
   1. Biology – Philosophy.   2. Essence (Philosophy)   I. Title.   II. Series.
   QH331.L29    2003
   570′.1 – dc21                                                    2003055148

ISBN  0 521 82599 7 hardback

# Contents

vii

# Preface

This book is rooted in my 1998 doctoral dissertation, written at the University of Massachusetts, Amherst: "Natural-Kind Term Reference and the Discovery of Essence." Roughly half of the chapters in the book (Chapters 2–4) have ancestors in the dissertation, though these chapters have all been extensively revised. Half of the chapters in the book (1, 5, and 6) are essentially new.

Because I outline the contents of this book in an Introduction, I will not do that here, too. Instead, I would like to express gratitude to many people and institutions for their support of this book. My dissertation director, Bruce Aune, is first on my list of those to whom I owe thanks. Bruce did much, both by criticism and by encouragement, to help to shepherd my dissertation to completion. I am very grateful for this philosophical and personal assistance and for the valuable help that Bruce has continued to provide since my graduate school days. Of particular relevance here is Bruce's trusted reaction to much of the newer material in this book. I owe much of whatever success the book enjoys to Bruce.

I am grateful to other members of my dissertation committee as well. Lynne Rudder Baker has offered highly valuable help and advice on the dissertation and much more. I am most grateful to her. Fred Feldman was a source of helpful feedback. I discussed some of the biology with James Walker, a biological systematist serving as the outside member of the committee. I also owe thanks to Lucy O'Brien, who was not on my dissertation committee. Lucy served as my M.A. thesis advisor at University College London, offering helpful feedback on early work concerning topics that I would later take up in the dissertation and then in Chapters 3 and 4 of this book.

I have received valuable feedback on work in this book from many other sources. There are too many people to list here, though some I have acknowledged in articles that have already been published. In addition to the anonymous referees for Cambridge University Press, the most salient sources of help were Phillip Bricker and Eric Schliesser, both of whom read earlier drafts of substantial portions of the manuscript and provided helpful and detailed comments. David Hull also provided valuable comments that led to significant improvements in Chapters 3 and 4.

Before I took a year's leave to work on the book in 2000–2001, I was pleased and fortunate to meet Michael Ruse when he visited Hope College. Michael suggested that when I finish the book, I submit it to the series for which he is the general editor, Cambridge Studies in Philosophy and Biology. It was a good fit, and I am grateful for the suggestion. Three of the first four foundational chapters of the book address whether a well-known philosophical tradition has application in the biological realm: This is a question for philosophers of biology. In the final two chapters of the book, I discuss consequences of claims made in the first four chapters. Although an emphasis on biology in these final two chapters is not required, it is appropriate: Biological examples allow for more continuity with the earlier foundational chapters than alternatives would, and biological examples illustrate the relevant lessons well.

As I indicate above, some of the work in the book has appeared in print already in the form of articles. Section (I) of Chapter 2 contains some material, reworked and reorganized, from my "Rigidity and Kind," *Philosophical Studies* 97: 293–316, copyright © 2000 Kluwer Academic Publishers, used with the kind permission of Kluwer Academic Publishers. Section (II) of Chapter 2 contains some material from "Essential Membership," *Philosophy of Science* 64: 96–112, copyright © 1997 by the University of Chicago. Chapter 4 contains some material from my 1996 article "Chemical Kind Term Reference and the Discovery of Essence," *Noûs* 30: 112–32. I gratefully acknowledge permission to use material from these journals.

Work on this book was generously supported by a grant from the National Endowment for the Humanities, an independent federal agency. Hope College generously provided time off from teaching and also further assistance, including but not limited to secretarial help on the bibliography and other matters: Nora Staal, Sally Smith, and especially Molly Baxter must be mentioned in this connection. With the NEH grant and the time off that it and Hope College provided, I was able to work on the book full time during the academic year 2000–2001. Had I not had that time to work, the book would not be finished today.

# Introduction

According to a celebrated philosophical tradition that has enjoyed prominence for more than a quarter of a century, so-called "natural-kind terms," like 'oak', 'water', and 'mammal', refer to kinds with theoretically interesting essences. According to the tradition, scientists learn by empirical investigation what those essences are. Scientifically informed conclusions about kinds' essences are discoveries, not stipulations. Scientists do not change the meaning of a term like 'water' or 'mammal'; they simply discover the essence of what speakers past and present have been calling "water" and "mammal," respectively.

The essence of water, scientists have discovered, is the chemical composition $H_2O$: Nothing could possibly be water without being $H_2O$ or $H_2O$ without being water, so long as scientists' empirical facts are right. If earlier speakers ever called something "water" that was composed of another chemical, they were wrong. In similar fashion, earlier speakers were wrong to call whales "fish." Scientists have corrected ordinary speakers about this matter. Empirical investigation into essences has led to the discovery that whales are mammals, not fish.

In this book, I examine the familiar tradition described above. The tradition is so well established that it is typically taken for granted in high-profile philosophical discussions from a wide range of diverse philosophical areas. Nevertheless, the tradition is mistaken. After showing that it is mistaken, I examine consequences. Here I provide a chapter-by-chapter overview of the book. Then I specify salient items not on my agenda.

## AN OVERVIEW OF THE BOOK

In Chapter 1, I clarify my subject matter. For the most part, I discuss biological taxa. Biological taxa include species, such as the lion (*Panthera leo*),

1

as well as higher taxa, which contain any number of species. Higher taxa that contain the lion species include the felines (a family) and the mammals (a class).[1] I argue that species and other taxa are natural kinds. I also defend the claim that various chemical substances are natural kinds, though my primary interest here, and indeed throughout this book, is with biological kinds. Some biologists and philosophers have argued that, contrary to the view I take, species and other taxa are not kinds at all but rather individuals. Other philosophers have argued that biological taxa and chemical substances are not natural. I argue otherwise.

In Chapter 2, I address essentialism, particularly with respect to biological kinds. Essentialists have not been well informed about biology, and I argue that some commonly held essentialist theses should be abandoned in light of contemporary biology. Nevertheless, I argue that some forms of essentialism about biological taxa are highly plausible in view of contemporary biological systematics and cladism in particular.

In Chapter 3, I address the further claim that biologists' conclusions about the essences of our kinds have been *discovered* to be true. Here I part with the familiar tradition, according to which scientists have not altered the use of key terms in sentences like 'Whales are mammals, not fish' to make such essence-revealing sentences true by stipulation; scientists have just found sentences like this to have been true all along. I argue that scientists' conclusions are not, in general, discovered to be true in this way. They are stipulated to be true.[2] Contrary to the received view, scientists change the meanings of kind terms.

In Chapter 4, I show that the lessons I draw in Chapter 3 about biological kinds apply more broadly. In particular, they apply to the other widely discussed group of natural kinds: chemical kinds. Chemists do not report the discovered essences of our chemical kinds any more than biologists report the discovered essences of our biological kinds.

The view that scientists' conclusions about the nature of our kinds are discoveries seems wrong. After showing this, I inquire into consequences. In Chapter 5, I address consequences for incommensurability. According to incommensurability theorists, rival scientists talk past one another. Words change their meanings as science progresses, so future theorists and past theorists who appear to disagree are really talking about different matters. Some friends of incommensurability go so far as to say that there can be no rational choice between competing theories. This threatens our ordinary, intuitive view of science's progress, according to which scientists advance by discarding older theories and replacing them with improvements.

The causal theory of reference to natural kinds is supposed to *redeem* the natural account of science's progress in the face of worries raised by incommensurability. That it redeems science's progress is supposed to be a major achievement of the causal theory. According to the causal theory, we do not define natural-kind terms by theoretical descriptions, which would be expected to change over time with our theories; otherwise the meanings of our kind terms would change with our theories, as incommensurability theorists say that they do. Rather, we identify samples of a kind, apply a name to the kind they exemplify, and then discover the essence of the kind. Theories can come and go without the reference of our natural-kind terms changing at all, for it is the samples in the world, and not our theory, that determine what our terms refer to. Thus, given the causal theory of reference, worries about incommensurability do not arise because worries about meaning instability do not arise.

Unfortunately, the foregoing account of reference stability is naïve. The meanings of our kind terms do shift, even given the causal theory, as I show in Chapters 3 and 4. Part of the problem may be that the causal theory does not completely discard descriptions, as I explain shortly. But even apart from descriptions and any instability they can cause, there is instability in reference. Often scientists find that it is not clear whether a term refers to this or that kind exemplified in samples: Empirical exploration uncovers many plausible candidates. At that point scientists often refine the use of the term in light of information that could not have been anticipated by the dubber of a term. Scientists change terms' meanings.

Thus, I argue that the causal theory does not rid us of meaning instability or ward off the threat that instability raises for the progress of science. I present a different defense of scientific progress. The main examples of progress upon which I focus are progress through the Darwinian revolution and progress through the rejection of vitalism.

The causal theory of reference cannot rescue us from problems of incommensurability, which threaten scientific progress. Neither can the causal theory rescue us from problems that threaten necessity, as I show in Chapter 6.

The new causal theory *appears* to liberate necessity from problems afflicting analyticity. Many acknowledge that Quine has cast serious doubts on analyticity, and in Quine's day, doubts about analyticity were doubts about necessity, also: Analyticity and necessity were not distinguished. Fortunately, in light of the causal theory of reference, it now appears that doubts about analyticity should not carry over to necessity. The causal theory allows for statements to be necessary yet *not* analytic, or even a priori. Thus, the many proponents of the necessity associated with the causal theory appear free to accept necessity and still concede that Quine is right about analyticity. So, given Quine's influence, it is not surprising that in the wake of the causal theory, there is little sympathy for analyticity: "Quine's rejection of analyticity still prevails," as William Lycan (1994, p. 249) and many others testify. But the causal theory, with its attending empirically discovered necessity, does not finesse objections to analyticity, as it is supposed to do. Chapters 3 and 4 strongly hint as much: My objections to essence discovery in those chapters are similar to Quineans' objections to analyticity. The suggestion is that empirically discovered necessity is subject to the same troubles as analyticity. If those troubles are so destructive that analyticity is completely untenable, then something similar ought to be suspected in the case of a posteriori necessity. And if a posteriori necessity is tenable, then analyticity's prospects brighten. Chapters 3 and 4 suggest this connection between analyticity and a posteriori necessity, but these chapters do not directly address analyticity.

In Chapter 6, I offer direct arguments that if causal theorists like Kripke are right about a posteriori necessity, Quine has to be wrong about analyticity. A commitment to the necessary a posteriori is a commitment to analyticity.

We must choose. We can have either the familiar necessity *or* the familiar animadversions to analyticity. We cannot have both, so one must be abandoned. I argue that it is the Quinean arguments against analyticity that should be abandoned. Chapters 3 and 4 may appear to commit me to choosing otherwise. They may appear to commit me to the familiar animadversions to analyticity. But I argue that they do not: It is not a consequence of my claims in Chapters 3 and 4 that analyticity is untenable.

So in the final two chapters of the book, I examine some consequences, real and apparent, of my attack on the view that scientists report the discovered essences of our natural kinds. There are many other possible consequences of interest.[3] But I leave other areas for other occasions.

Some projects that might be *thought* to be part of my agenda are not part of it. In particular, it might appear that I aim to undermine the causal theory of reference and to replace it with an old-fashioned descriptivist account. There is a well-known problem that threatens the causal theory, and this problem bears some resemblance to considerations that I raise in Chapters 3 and 4. But I do not think that this problem, which is known as the "qua problem," discredits the causal theory. In a few brief paragraphs I will explain why I accept some form of the causal theory, and why I think that the theory survives the qua problem. This provides me with an occasion to introduce and clarify the causal theory, which is discussed throughout most of the book, as well as to forestall a possible misunderstanding about my aim.

*Introduction*

The causal theory addresses how a word is assigned to its referent by the word's dubber. It also addresses how a word is passed on from the dubber to other speakers: With respect to this latter issue, the causal theory recognizes a chain of communication, but this is of little importance in what follows. The issue of how a word is assigned to its referent is more important. According to the causal theory, a term for an individual is typically coined in the presence of the individual by a formal or informal baptismal ceremony in which the speaker says something like "This person is to be called 'Cicero' " (the speaker points). A term for a natural kind is typically baptized in the presence of *samples* of the kind: "*Tyrannosaurus rex* is the species instantiated by *that* fossilized organism." The causal theory is typically contrasted with an old-fashioned descriptivist account, according to which a kind term refers to something just in case the candidate satisfies descriptions that speakers can list without reference to a sample, such as 'carnivorous', 'reached thirty-nine feet in length and weighed ten tons', 'had two puny forelimbs and two powerful hind limbs', 'had dagger-like teeth', 'lived some seventy million years ago in North America', and so on.

The causal account seems to capture the typical naming practices of biologists. Biologists coin new species terms by providing a *sample*, called a "type specimen." The newly coined term refers to the species instantiated by the type specimen. The type specimen is generally stored in a public location in order to allow biologists to use it for reference. Uniquely identifying descriptions that make no reference to a type specimen are not needed and are often not used. Consider a simple example. Two biologists in the Amazonian rain forest today coin new species terms. Each collects a type specimen, specifies a term, and says, "Let this term name the biological species exemplified in this organism" (the speaker points). Suppose that the two type specimens belong to different biological species: They are not related by evolution, would be incapable of interbreeding, and so on. Then even though the two biologists may have failed to offer uniquely identifying descriptions of their species, the biologists have still coined terms for *two different biological species*. Further investigation could determine this to every biologist's satisfaction.

A uniquely fitting description that makes no reference to the sample is not needed in order to assign a species term to a species. Further, it may not even be available. The type specimen may consist of only a fragment of a skeleton preserved in a fossil, so that the biologist does not have a very clear idea what the species was like. Or the type specimen may be an atypical member of the species (see, e.g., Ereshefsky 2001, p. 261). For these reasons, the type specimen may not provide enough information to allow the species term dubber to give a uniquely identifying description that makes no reference to the specimen.

Old-fashioned descriptivism seems untenable. Paradigmatic samples are used in reference.[4] That is not to say that descriptive information is irrelevant for reference according to the causal theory. Baptisms convey some descriptive information about the referent. When a dubber points and says, "Let this term name the biological species exemplified in this organism," she specifies that the referent is a *species*.

More descriptive information may be offered. A *reference-fixing description* that appeals to a cause may be used to baptize a term and may even be used in place of pointing. For example, one might coin a species term '*S*' by saying "*S* is the species that left footprints in such and such location."[5] Here the description fixes the reference of '*S*', as opposed to serving as a synonym for '*S*'. As it happens, prehistoric artiodactyls may have left footprints in the location, so that the relevant species of artiodactyl is the referent of '*S*'. But it is hardly a necessary truth that this species of artiodactyl, *S*, left prints in that spot. Some predatory species of cat could have settled on the spot, leaving its own prints instead and chasing away all artiodactyls, including specimens of *S*. If 'the species that left footprints in such and such location' were a synonym for '*S*', rather than simply a description to fix the reference, then '*S* is the species that left prints in such and such location' would be a necessarily true statement; instead, it is a contingently true statement.

'Causal theory' should not suggest, then, that no descriptive information is relevant to a term's reference.[6] That descriptive information is relevant for the causal theory helps that theory to avoid various objections, including one that might be mistaken for mine: the objection that the theory suffers from the so-called "qua problem." This problem may be illustrated as follows. Take a term like '*Tyrannosaurus rex*'. Causal theorists would say that the term is "grounded" in a sample *T. rex* specimen that has been fossilized. But if that is so, why does '*T. rex*' apply to *T. rex* instead of to dinosaurs in general or to animals in general or to big things in general? The sample was, after all, a dinosaur and an animal and a big thing as well as a specimen of *T. rex*. The causal theory seems, at least on first blush, to be unable to account for the reference of '*T. rex*' to *T. rex* instead of to these other kinds. It seems to be undermined by a qua problem.

This problem is not fatal. Recall the baptism specified above: "*Tyrannosaurus rex* is the species instantiated by *that* fossilized organism." Given such a baptism, the term '*T. rex*' *has* to refer to *T. rex* instead of to the kinds *dinosaur*, *animal*, or *big thing*, for the simple reason that *T. rex* is the only kind in the group that is a *species*. The dinosaurs as a whole, in all of their diversity, do not constitute a single species, though they constitute a larger taxon (see note 1). Similar words apply to the animals in general. The kind *big thing* is obviously not a species. So if '*T. rex*' was coined as specified above, then it had to be assigned to *T. rex*, as opposed to kinds like these others that also happen to be instantiated by *T. rex* specimens. And '*T. rex*' *was* coined pretty much as specified above. Early in the twentieth century, Henry Fairfield Osborn (1905) published the name '*T. rex*' for the first time, explicitly specifying that it was a name for the *species* exemplified in a certain fossilized specimen.

For this and related reasons, the qua problem is not fatal to the causal theory. That is not to say that the causal theory eliminates indeterminacy. But eliminating indeterminacy is not its job. Its job is to recognize just the indeterminacy that is really in the language, and not more than that. If there is a qua problem, then the causal theory recognizes far too *much* indeterminacy. This would be a problem, but it is not a problem that the theory recognizes any indeterminacy at all. The causal theory does not recognize too much indeterminacy, because some descriptive information can be used in causal baptisms.

Other problems are sometimes thought to afflict the causal theory and seem to admit a similar resolution, but I do not discuss them here. A comprehensive defense of the causal theory is not in my plans. Here I wish simply to affirm that I endorse some version of the causal theory and to distance myself from the foregoing criticism in particular. Although I argue that according to the causal theory or any plausible theory there is often a good deal of interesting indeterminacy in the reference of our terms, this is not a criticism of the causal theory. It is a recognition of indeterminacy in the language. I will not be arguing that causal theorists are committed to recognizing any indeterminacy that is *not* present in the language, as the qua objection does.

Although my aim is not to criticize the broad outlines of the causal theory, I offer substantial refinements. More importantly, I suggest that the causal theory does not alter the philosophical landscape nearly as much as it has been supposed to do. In particular, the theory does not clear up or circumvent famously troubled issues that concern referential instability and that concern necessity because of its connection to analyticity. Despite appearances and despite much testimony to the contrary, these big, fascinating issues about conceptual change are not much affected by the advent of the causal theory.

# 1

# What Is a Natural Kind,
# and Do Biological Taxa Qualify?

Before I undertake an examination of natural kinds and reference to them, I will clarify what natural kinds and natural-kind terms are. Philosophers who discuss natural-kind terms seldom offer any analysis of what they are supposed to be. Prominent philosophers tend just to offer examples. Their examples are supposed to help readers grasp intuitively what is intended. The most common examples of natural-kind terms presented in the literature are perhaps terms for biological species and higher taxa: 'Tiger', 'elm', and 'mammal' are all discussed extensively in the literature. Chemical kind terms, such as 'water' and 'jade', are also presented as examples with great frequency.

My primary interest in this book has to do with kind terms from biology. To a lesser extent I also examine the other prominent group of natural-kind terms, those from chemistry, in order to show that many of the central observations that I make about specifically biological kinds are not limited to the realm of biological kinds.

In this chapter, I defend the position that the terms from biology and chemistry that are typically supposed to be natural-kind terms really are natural-kind terms. As I have said, most analytic philosophers accept terms like 'tiger' and 'water' as paradigmatic, as words that are "natural-kind words if any are," as one introductory textbook says of them (Platts 1997, p. 264). Biological species in particular tend to be regarded as paradigmatic natural kinds. "The classification of living things into biological species is one of the paradigm (i.e. indisputable and/or typical) cases of a division into 'natural kinds'," another textbook says (Wolfram 1989, p. 236).

Unfortunately, various obstacles threaten to create problems for textbook examples of natural kinds. Vigorous criticisms have been launched at the dominant perspective from both philosophers and biologists. Some critics charge that many or all of the supposed biological natural kinds are not natural kinds because they are not really *kinds*. Instead, species and perhaps even higher taxa are individuals. Other critics charge that the supposed biological natural kinds and even chemical kinds are not natural kinds because they are not really *natural*. In my view, both criticisms miss the mark. I defend the position that biological taxa, as well as chemical substances, are natural kinds.

## I. ARE BIOLOGICAL TAXA KINDS?

The first objection to counting the usual list of natural kinds, or rather the usual list of natural kinds from biology, as a list of genuine natural kinds is that the items on the list are not *kinds*. The objection is that species are individuals rather than kinds (Ghiselin 1974, 1987, 1995, 1997; Hull 1976, 1978, 1980, 1987). Perhaps higher taxa too, are individuals instead of kinds (Ghiselin 1995; see also Ereshefsky 1991 and references therein). Being individuals, species are concrete entities like Mt. Rainier or George W. Bush. They differ markedly from *gold* or *water*, which are kinds and therefore abstract entities.[1]

In the present section (I), I defend the claim that species are kinds. In sections (I.1)–(I.5) I find fault with the usual reasons for saying that species are *not* kinds. In section (I.6) I offer a positive argument for the claim that species *are* kinds and discuss the rival view that they are instead individuals.

### I.1. *Evolving Kinds*

The most frequently cited reason for supposing species to be individuals is that they evolve. "If species were not individuals, they could not evolve," Ghiselin (1987, p. 129) insists. Natural kinds, in particular, cannot evolve. Ghiselin sees this as "the most compelling" reason for counting species to be individuals rather than kinds (1981, p. 303). Kinds are deemed incapable of evolving because they are abstract objects, with immutable essences. A natural kind cannot change in any respect. Only particular concrete objects are capable of change. So if species evolve, they must be individuals, not kinds.

This objection is not persuasive. Abstract objects' incapacity for change is certainly a barrier to their evolving, but when we say "Species evolve," we do not mean that any abstract kind evolves; we mean that successive *members* of a kind gradually become different from their ancestors. Similarly, by the claim "One species can evolve into another" what is communicated is not that any abstract object can become a different abstract object but rather that the instances of one species-kind can give rise to instances of another species-kind.

A parallel can be drawn to incontrovertible kinds such as lead and water. Elements like lead were generated from lighter matter. There is nothing paradoxical about this claim. What is communicated by the statement 'Lead has been generated from lighter matter' is not that an abstract kind, lead, has been generated; what is communicated is rather that concrete *members* of the kind came into being by the above generative process.[2] In the same way, when 'Lead can be transmuted into gold' is correctly affirmed, the intent is not that one abstract kind is transmuted into another but rather that lead particles can be transmuted into gold ones. Again, when speakers affirm the statement 'Water can be created with hydrogen and oxygen', what they communicate is not that two abstract kinds can be used to create a third abstract kind. Rather, what they communicate is that instances of two abstract kinds can be used to create instances of a third abstract kind. In none of these affirmations about what are clearly kinds is there any commitment to abstract objects' changing. The same holds for the statement 'Some species of fish evolved into a species of amphibian'. What is communicated by that statement is not that one abstract object changed into another but that instances of the one kind generated, by a process of evolution, instances of another. The argument that species cannot be kinds because they evolve fails for taking idioms too seriously.

## I.2.  *Historical Kinds*

Another common objection to counting biological taxa as kinds alleges that species and higher taxa are spatio-temporally restricted, whereas kinds are spatio-temporally unrestricted. No matter how similar to our terrestrial horses Alpha Centaurian organisms may be, they are not members of the horse species. Genetic similarity, interfertility, and so on could not establish conspecificity, because the Alpha Centaurians are not historically connected to the terrestrial species. Therefore, the horse species is not a kind. Hull draws a comparison to the undisputed kind gold:

> If all atoms with atomic number 79 ceased to exist, gold would cease to exist, although a slot would remain open in the periodic table. Later when atoms with the appropriate atomic number were generated, they would be atoms of gold regardless of their origins. But in the typical case, to *be* a horse one must be *born* of horse. (Hull 1978, p. 349; see also, e.g., Ghiselin 1981, p. 304)

Hull notes that once a species has disappeared, the impossibility of reappearance is "conceptual" rather than contingent. And this is supposed to reveal that species cannot be kinds, being historically delimited, but rather that they must be individuals.

This is not persuasive. Historically delimited species can be kinds. There is evidently no reason there cannot be historical kinds as well as nonhistorical ones. A historical kind would simply be one whose membership conditions involve members' having some causal connection to an independently specified item – for example, the beginning of a lineage.[3] If *horse* or *oak* is spatio-temporally restricted, then to be a member of the horse kind or the oak kind, an object would need to be an organism of this or that lineage. I will have much more to say about historical kinds in the sections that follow and in the next chapter.

### I.3. *Species' Essences*

If species must be historical, as individuality theorists and a growing majority of biological systematists and philosophers of biology say, then the view, popular in the philosophical community, that each species is identical to a genetic kind is mistaken. An Alpha Centaurian animal genetically identical to the horse would belong to a different species, so the horse kind is not identical to the relevant genetic structure. No genetic code captures the essence of the species.[4]

Ghiselin insists that species not only fail to have *genetic* essences, but they also fail to have *any* essences or essential properties at all. "That species are individuals means that they cannot be defined, in the sense of listing properties they simply must have," though of course they can be described as having certain contingent properties (1987, p. 129). The same holds for *clades*, a clade being an ancestral group or "stem," and all of its descendants.[5] Cladists, who belong to the increasingly dominant cladistic school of systematics, identify higher taxa, like Mammalia (the mammals), Aves (the birds), or Serpentes (the snakes), with clades.

In places Ghiselin suggests that the absence of essential properties in species and clades supports his thesis that species and clades are individuals (see, e.g., 1981, p. 304). But Ghiselin is wrong in saying that species and clades have no essential properties. Clades *do* have essential properties, as Kevin de Queiroz (1995) suggests. Take the clade that de Queiroz proposes to call "Mammalia" (1995, pp. 224–5). As de Queiroz defines 'Mammalia', the term applies, in actual and counterfactual circumstances, to the clade that is descended from a particular stem. The stem is the most recent ancestor common to both the horse and the echidna.

Ghiselin has pointed out that it is contingent, not necessary, that Mammalia ever gave rise to horses or echidnas (1995, pp. 220–1). He is right about this, but his observation is not damaging for essentialism. To make the essentialist lesson clear, I propose to *name* that group that happens, as a matter of contingent fact, to be the most recent ancestor common to both the horse and the echidna. I give it the name '*G*'. A cataclysm could have wiped out *G* before it ever gave rise to the horse or the echidna, so it is not necessary that *G* gave rise to the horse and the echidna. But although it is contingent, not necessary, that *G* gave rise to the horse and the echidna, it is *necessary* that any organism descended from *G* belong to the clade Mammalia, and that any organism belonging to the clade Mammalia be descended from *G*. Mammalia therefore has some necessary properties. It is necessarily such that it includes *all* organisms in *G* or descended from *G*. And it is necessarily such that it includes *only* organisms in *G* or descended from *G*.

There are many other examples of clade names that have been similarly defined. Willi Hennig, the founder of cladism, suggests a similar definition for 'Aves', the scientific name for the birds. Hennig proposes that the "stem" of the birds, or Aves, is the species that gave rise to all of the known species of birds. Hennig supposes that this is probably some species of *Archaeopteryx* (Hennig 1966, p. 72). Again, I propose a name for that species: call it "*A*," though the name is to apply even if the bearer turns out to be something other than a species of *Archaeopteryx*. The relevant clade includes, in any possible world, all and only the organisms in and descended from species *A*.

Mammalia and Aves are higher-level groups that contain many species. Similar reasoning reveals that *species* have essential properties on standard historical conceptions. Species, such as the horse species *Equus caballus* or the ostrich species *Struthio camelus*, necessarily stem from their ancestral groups just as clades do, though a species differs from a clade in that its descendants do not all have to belong to it. I return to the topic of essentialism with respect to species and clades in Chapter 2.[6]

## I.4.  *Ostension*

Hand in hand with Ghiselin's conviction that species and clades have no essential properties is his view that reference to them is achieved only by the use of ostension: "Individuals have to be defined 'ostensively' – by 'pointing' – for they have no defining properties whatsoever" (1987, p. 134). That we point to achieve reference to species and clades provides reason to take them to be individuals, Ghiselin suggests. "Even the very act of 'pointing' at a person or a common ancestor is the sort of particular action that has the sort of concreteness that characterizes individuals rather than classes" (Ghiselin 1995, p. 222). That we achieve reference to species and clades by ostension is one of the major reasons Ghiselin became convinced that species are individuals (Ghiselin 1999, p. 449).

I have already argued (in §I.3) that species and clades do have essential properties. Terms naming them might be coined by appeal to those properties, without any pointing. One can define the clade that de Queiroz calls "Mammalia" as follows: "Mammalia $=_{df}$. the clade whose stem is group $G$." In that case *having $G$ for a stem* is an essential, defining property of the clade. But no one has pointed at $G$ in order to define the clade's name. So pointing at a common ancestor is not necessary for defining the names of clades, contrary to Ghiselin.

Ostension *could* be used to secure reference to a taxon, it is true. But if it is, this does not suggest that the taxon is an individual rather than a kind. Speakers may use ostension when securing reference to kinds too, at least if something like the method of term baptism recognized by Kripke and Putnam is coherent. No one may know the underlying composition of the substance in a particular vial, though a term is desired for the compound or element in the vial. In that case, speakers could coin a term by saying, "This term is to refer to the chemical compound or element instantiated by the stuff in this vial" (the speaker points). This might, if the sample is homogeneous, result in the term's referring to $H_2O$ or $NaCl$.[7] So a baptism that makes use of pointing is no mark of a term that refers to an individual. Even if names for taxa are baptized by ostension, nothing much follows that would threaten the thesis that taxa are kinds.

## I.5. *Laws About Particular Species*

An alleged implication of the s-a-i (species-as-individuals) thesis is that there can be no laws about particular species. Thus, the s-a-i thesis is often said to explain the absence of laws about species: The reason there are no laws about particular species or other taxa is that *there can be no laws about individuals*. Laws apply only to classes or kinds.

Because it appears to explain the absence of laws about particular taxa, the s-a-i thesis has won some eminent converts, including Hull (1987, pp. 173–4, 183). As Ghiselin explains,

> What convinced Hull was the fact that there are no laws of nature for particular species. This made perfectly good sense because there are no laws

of nature for any individuals whatsoever. Laws of nature, which are necessarily true of everything to which they apply, irrespective of time and place, are about classes, such as "planet" or "species" in general, and make no reference to individuals such as "Jupiter" or "*Homo sapiens.*" (1999, p. 449)

I am unconvinced that laws of nature and their boundary conditions must "not include reference to particular times and places" (Hull 1987, p. 174), as Ghiselin and Hull suppose. It is certainly not evident from the very nature of a law that no law could refer to a particular time or place, as Marc Lange observes:

[R]ecall P. A. M. Dirac's famous conjecture that the gravitational-force "constant" is inversely proportional to the time since the Big Bang, which would require that a statement of the gravitational-force law include an implicit reference to some particular moment (for example, refer to the time elapsed since the Big Bang, which Dirac calls "a natural origin of time"). (1995, p. 433)

Dirac's conjecture may be false, but it is hardly apparent that because it refers to a particular time it *could* not have been a law. Similarly, it seems wrong to say that laws could not possibly refer to individuals, that "laws have to be generalizations about classes of individuals" (Ghiselin 1989, p. 53). If by the very nature of a law, no law could make reference to an individual, then Aristotelian physics, say, according to which some law statements refer to the path of the moon, could be rejected without a look at the empirical world, by anyone who understands the nature of a law statement. That seems wrong (see Lange 1995, p. 433; Armstrong 1983, p. 26; Tooley 1977, p. 686).

It is not clear that there can be no laws about individuals, as s-a-i theorists have supposed in promoting their thesis. Furthermore, even if it is *true* that there can be no laws about individuals, it does not *follow* that species are individuals. If there can be no laws about individuals and if species are individuals, then, of course, it does follow that there are no laws about species. But if there can be no laws about individuals and if there are no laws about species, it does not follow that species are individuals. Species could still be lawless kinds. I argue in the following section that there are good reasons for taking species to be kinds, but species could be lawless kinds for all that I claim. My arguments that species are kinds are compatible with there being no laws about any species.

## I.6.  *Kinds vs. Individuals*

I have now countered the most prominent and frequently repeated reasons for supposing that species and other taxa are not kinds. But I have not argued that species and other taxa *are* kinds. Nor have I ruled out the rival position that they are individuals instead of kinds. It is time to address these tasks.

I.6.a.  SPECIES ARE KINDS. Before I address directly the question of whether species are individuals, I will argue that species are kinds. Perhaps an argument that species are kinds is superfluous, at this point, because that position seems to be the default position. That species are kinds seems the most natural position to take, in the absence of better arguments to the contrary than opponents seem to have provided. The kindhood of species is suggested by the similarity between kinds like *water* and *gold*, on the one hand, and apparent biological kinds like *radish* and *tiger* on the other. Both seem to have members, namely individual bits of water and gold, on the one hand, and individual radishes and tigers (or parts of radishes and tigers), on the other. The usual practice of treating all these entities alike as kinds would seem to be reasonable in the absence of good objections, and as I have said, there do not seem to be any good objections.

But more can be said to motivate the position. A kind is closely related to a property. Some authors say that a kind *is* a property of some sort,[8] but whether the relationship is that close does not matter much to what follows. What is important here is that for any property, there is a corresponding kind the essential mark of which is to possess that property. This seems evident. For example, to the property of being red there corresponds redkind. Redkind is a kind, though perhaps not a natural kind, the essential mark of which is possession of redness: The members of redkind in any possible world and at any time are just the red objects in that world at that time.

Suppose that the organisms of any species make up an individual, or something else that is not a kind. Call such an object a *species-individual*. Suppose, further, that talk about the species could satisfactorily be interpreted as talk about the species-individual. In that case, I will argue, such talk about the species could *also* be satisfactorily interpreted as talk about a kind. Here is why: Although the species-individual is not a kind but rather an individual, there is a property, for any such individual, of being part of that individual. For that property, just as for any other property, there is a corresponding kind, such that possession of the property is the essential mark of the *kind*. Thus, for any species, there is a kind the members of which are precisely the parts of its species-individual. Call such a kind a *species-kind*. Because, for any species,

15

there is both a species-individual and a species-kind, there is no reason that discourse about the species would have to be interpreted as being about the species-individual rather than the species-kind.

An example may help to clarify the reasoning. The organisms belonging to *Raphanus sativus* (and their parts) are precisely the parts of a species-individual. Those parts are precisely the members of a species-kind. Thus the organisms belonging to *Raphanus sativus* are precisely the members of a species-kind. And it would therefore seem that discourse about the species *Raphanus sativus* could satisfactorily be interpreted as discourse about that species-kind, whose members are just the organisms of *Raphanus sativus*.[9]

It should be clear from the foregoing discussion that species-kinds do not differ from species-individuals by being any less scientifically interesting. So my account of species-kinds differs vastly from that of Hey (2001), for example: For Hey, there may be but need not be any correspondence between the parts of a "real species," which is an individual evolutionary group, and the members of any *kind*, which is a group of relatively superficially similar items (Hey 2001, pp. 105ff.; for related thoughts, see Dupré 2001, pp. 204–5). On my account, by contrast, a species-kind is not a group of superficially similar organisms any more than a species-individual is. If a species-individual's parts form an evolutionary group, the corresponding species-kind must have as its members not organisms that are similar in something like appearance but rather the very same organisms that serve as parts of the species-individual: Thus, the *kind*'s members form an evolutionary group, just as the individual's parts do. That is why talk about the relevant species can be interpreted as being about a species-kind, whose members serve as an evolutionary group, as readily as such talk can be interpreted as being about a species-individual, whose parts serve as an evolutionary group.

I have argued that species are kinds. More precisely, I have argued that talk about species can be interpreted as talk about kinds. I have not yet addressed higher taxa and clades (higher taxa are now generally understood to *be* clades). Although it is sometimes thought to be more problematic to say that clades are kinds than to say that species are kinds (Dupré 2001, pp. 204–5), the foregoing arguments concerning species apply readily to talk about clades. Any clade's constituent organisms constitute an individual or some other sort of concrete entity: Call this a "clade-individual." There is a property of being a part of that individual. The objects with that property are precisely the members of a kind: Call this a "clade-kind." Because the parts of the clade-individual are identical to the members of the clade-kind, talk about a clade may be interpreted as talk about a kind.

I.6.b. ARE SPECIES INDIVIDUALS? I have argued that there are species-kinds, and that discourse about species can be interpreted as discourse about species-kinds. What is to be said, then, for the claim that species and other taxa are individuals? As it turns out, there is no need to refute that claim, properly understood. There are well-known objections to the position that species are individuals – for example, that species are not integrated enough to be counted as individuals (Mayr 1987, pp. 158–61; Ruse 1987, pp. 232–5; Borjesson 1999, pp. 879–96). Though such objections are interesting, there is no need to evaluate them here. Let it be *granted* to s-a-i theorists that the organisms of a species constitute an individual, either because they display cohesion after all or because individuality, or individuality of the relevant stripe, does not require cohesion (Wilson 1999, pp. 83–4). Even if it is granted in this way that there is an individual whose parts are the organisms of a species, it is nevertheless the case that there is a *kind* here, as well. Given both a kind and an individual, it seems plausible to suppose that fans of part-talk will prefer in general to reconstruct species-talk as talk about the parts of an individual, and fans of membership-talk will prefer to reconstruct species-talk as talk about kinds (sets, whatever).[10] Because talk about species is not clearly about individuals as opposed to kinds or vice-versa, either interpretation involves refinement, but the refinement will not affect the acceptability of ordinary scientific claims. Either reconstruction can make satisfactory sense of species-talk.

There is, then, no need to reject the idea that the organisms of a species constitute an individual. I have argued only that there are species-kinds, and that it is reasonable to interpret talk about species as talk about those kinds. I discuss species-kinds, and kinds corresponding to higher taxa, in the pages that follow. It remains to be discussed whether these entities are natural, but they are kinds.

## II. ARE BIOLOGICAL TAXA AND OTHER KINDS FROM OUTSIDE PHYSICS NATURAL?

In the remainder of the present chapter, I argue that biological taxa, as well as chemical kinds, are natural. In the first part of the present section, part (II.1), I inquire into what it is for a kind to be natural, and I argue that biological taxa and chemical kinds meet the requirements. In section (II.2), I argue that various kinds are natural in different respects and to different degrees. In section (II.3), I argue that what is appropriately called "natural" varies with the context. This provides me with a response to skeptics who say that my list of natural kinds should be much narrower than it is, and that biological kinds are not natural. It also provides me with a response, as I show in section (II.4), to critics who would claim that my list of natural kinds should be much broader than it is. In section (II.5), I address the issue of whether the vernacular is a source of natural-kind terms.

## II.1.   *What Is a Natural Kind?*

What is it that distinguishes *natural* kinds from artificial kinds? This is not obvious. An initially plausible proposal is that natural kinds are kinds found in nature. On this account *tiger*, *elm*, and *water* all qualify as natural kinds. *Toothpaste*, *lawyer*, and *trash*, on the other hand, fail to qualify as natural kinds. So the proposal succeeds in distinguishing these kinds in an intuitive way. Nevertheless, this proposal seems to stumble over other examples. Not all human-made kinds fail to be natural kinds. Humans have produced minerals, such as quartz and diamond, in the lab. Humans have also produced elements. Technetium is a synthetically produced element that has not been found to occur naturally on Earth.[11] And humans have created new species of plants by inducing polyploidy.

Not only are not all natural kinds produced in nature, but not all kinds in nature are natural kinds: Consider *mud*, *dust*, or *shrub*. These are too close to the toothpaste and trash kinds to count as natural. Natural kinds are not distinguished by being found in nature.

Rather than try to illuminate natural kinds, one might, of course, try to find a satisfactory characterization of natural-kind *terms*. An initially plausible suggestion for identifying natural-kind terms is to say that they are those terms whose extensions are determined by experts. If you have a question about whether you are right to call the white spots at the back of your throat "streptococci," or to call a tree in your yard an "elm," you consult someone who knows more than you do about these matters. There is, as Putnam (1975e) puts it, a division of linguistic labor.

But the division of linguistic labor's applicability is not what distinguishes natural-kind terms from other terms. 'Glass', for example, is a dubious candidate for being a natural-kind term, because unlike, say, diamond, glass may be made of different kinds of chemicals. Even so, an ordinary person in doubt about the composition of a bit of glass-like material would have to consult an expert to find out whether it is really glass. Furthermore, a term that is a natural-kind term could fail to be marked by any division of linguistic labor, as Platts (1997, p. 267) observes. Its use might be limited to a small community of experts, so that no one in the community relies on others for help in determining the word's extension.

Yet another suggestion might be that a natural-kind term is one whose conditions for application are determined by the nature of paradigmatic samples in the world. This nature must be empirically determined.

A problem with this characterization is that even if a term is baptized with the use of samples, as many natural-kind terms are, it still might turn out *not* to refer to a natural kind. 'Reptile', for example, was coined as a natural-kind term, but many systematists insist that the reptiles do not constitute a natural group (see Chapter 3 for details). They constitute an artificial group. Empirical investigation into the samples was needed to determine that, so the need for such empirical investigation is no mark of natural-kind terms. A similar term from chemistry might be 'air', which turned out to designate a motley assortment of chemicals. A term is a natural-kind term not in virtue of being *coined* to refer to a natural kind but in virtue of *really* referring to a natural kind.

Another problem with saying that a natural-kind term is just one that is grounded by samples in the world is that some natural-kind terms seem to be defined by theory rather than coined in a baptismal ceremony using samples. 'Water' was plausibly coined with the help of samples, but '$H_2O$' seems defined by theory, not samples. Yet '$H_2O$' seems to be a natural-kind term.

The real distinction between natural and non-natural kinds seems to have to do not with where they are found or how people manage to get words to refer to them but with their theoretical significance. In particular, I propose, a natural kind is a kind with explanatory value. A lot is explained by an object's being a polar bear. That it is a polar bear explains why it raises cubs as it does, or why it has extremely dense fur, or why it swims long distances through icy water in search of ice floes. Similarly, there are theoretically satisfying answers as to why polar bears on the whole raise cubs as they do, or have dense fur, or swim for miles through icy water. The polar bear kind is a useful one for providing significant explanations. It is a natural kind.

Compare a highly unnatural kind, such as the *named-on-a-Tuesday* kind, to the polar bear kind. The named-on-a-Tuesday kind has as members all and only objects named on a Tuesday. Some people, some animals, some machines, and some planets belong to the named-on-a-Tuesday kind, but almost nothing about them is explained by their membership in the named-on-a-Tuesday kind. Nor are there any interesting explanations about the named-on-a-Tuesday kind in general. The named-on-a-Tuesday kind is entirely unlike the polar bear kind. It is not a natural kind.[12]

Kinds like toothpaste or trash are also unlike the polar bear kind. While not as artificial as the named-on-a-Tuesday kind, toothpaste and trash do not measure up to the naturalness of the polar bear kind. Naturalness is not all or nothing, as these examples indicate. I must say more about this.

## II.2. *Naturalness in Respects and Degrees*

I have proposed that the naturalness of a kind consists in its explanatory value. Because natural kinds have explanatory value and are, at least in the best of circumstances, useful for prediction and control as well, they provide the basis of scientific classification (Daly 1998, p. 683). In biological classification, the first aim is to reflect history. As Hull emphasizes, taxa are supposed to be pieces of the "phylogenetic tree. The primary goal of taxonomy since Darwin has been to reflect . . . successive splittings in a hierarchical classification made up of species, genera, families, and so on" (Hull 1998, p. 272).[13] Because taxa divide the tree of life into historical units, they are beneficial for historical explanation. We can explain, for example, the relationships between the polar bear, the brown bear, and the black bear in terms of genealogy. Similarly, we can answer why polar bears are capable of swimming for miles in icy water by appealing to a historical explanation something like this: Some animals in the population of bears ancestral to those living today were better able to survive than others, because they could better swim to new seal-laden floes. These passed on the traits that make for efficient swimming in these conditions.

The foregoing type of explanation indicates that the polar bear kind is not only interesting from a historical point of view. Polar bears have a common anatomy and similar behavior, they occupy together an ecological niche, and so on. Therefore the group can find a place in anatomical explanations, behavioral explanations, ecological explanations, and so on, as well as evolutionary explanations. This makes the polar bear kind *more* natural than a group with little in common except a shared history.

Sometimes groups that are natural from a genealogical perspective are less natural from another perspective proper to biology. This gives rise to disagreements in classification. Any taxon recognized by systematists of the cladistic school includes every descendant belonging to any organism in the taxon. Taxa recognized by evolutionary taxonomists, on the other hand, sometimes exclude some descendants of organisms belonging to the taxa. Neither cladists nor evolutionary taxonomists recognize groups that are unrelated historically, so both cladistic and evolutionary taxa may take part in evolutionary biologists' explanations. But cladistic taxa are distinguished on the basis of pure genealogy. Evolutionary taxonomists take into account other kinds of information too, like ecological information or genetic information.

An example over which cladists and evolutionary taxonomists disagree can illuminate differences. As an evolutionary taxonomist, Mayr (1995, p. 431) recognizes the barbets (Capitonidae) and toucans (Ramphastidae), two groups of tropical birds. The barbets are spread out from Asia to Africa and South America, but they are everywhere very similar, and everywhere they fill the same adaptive zone. The South American barbets have given rise to descendants, the toucans, which are strikingly dissimilar from the various barbets, and which occupy a very different adaptive zone. Evolutionary taxonomists group all of the barbets into one family but classify the toucans in a separate family. Because the family of barbets fails to include all descendants of the organisms belonging to the family (it fails to include the toucans), cladists do not recognize this family. Cladists combine the toucans with their South American barbet ancestors into one family and the barbets from Africa and Asia into another. This grouping is more informative with respect to genealogy but less informative with respect to ecology. The tradeoff in information is clear. Thus, Mayr says,

> The Hennigian [cladistic] reference system and the traditional [evolutionary] classification are specially suited for different objectives. If mere genealogy, mere line of descent, is the information that is wanted, the Hennigian system is superior. However, if the student wants relatively homogeneous taxa, largely based on similarity and on the degree of genetic relationship, also reflecting their niche occupation, a traditional evolutionary classification is preferable. (1995, p. 431)

Mayr has some grounds, therefore, for resisting cladists' accusation that the traditional taxa he favors are "artificial" (1995, p. 429) or "arbitrary" (p. 425). The barbet family is more natural than the cladistic taxon comprising the South American barbets plus toucans in the respect that the barbet family shares an adaptive zone, among other things.[14] But the barbet family is less natural than cladistic counterparts in the respect that it does not do as good a job reflecting genealogy. It is like a group consisting of a nuclear family minus one or two of several children.

Fortunately, cladistic and evolutionary taxonomists' classifications recognize many of the same groups (Mayr 1995, pp. 430–1). In general a common history is a good indication of other shared traits. Genealogy is sometimes indiscernible, but a purely genealogical classification, if one could get it, would be likely to place organisms into groups that are natural in a number of respects. As Darwin says, a genealogical classification would be most useful "if we had a real pedigree," because "the principle of inheritance would keep the forms together which were allied in the greatest number of points"

21

(1859, p. 423). This consilience provides motivation for systematists' continued struggle to discern phylogeny.[15]

When different systems of classification agree, that is because the taxon in question is natural in a number of respects. A taxon that is recognized by both cladists and evolutionary taxonomists is a natural genealogical group *and* a natural ecological group. It is therefore more natural, other things being equal, than, say, a group that enjoys a common genealogy but not a common ecology. A group that is natural in a number of respects is valuable for many kinds of explanation. There is a close connection between naturalness and consilience, as Ruse (1987, pp. 237–9), among others, emphasizes.[16]

I have claimed that taxa can be natural in different respects. They can also be natural in a particular respect to different degrees. Groups narrower than those groups that taxonomists address may be *more* natural for some purposes than taxa. It is to be hoped, for example, that scientists will continue to identify narrow kinds that are more natural than species with respect to particular medical inquiries. The success of the human genome project suggests that this will be the case. Rather than having to rely on generalizations about the human body, future physicians will, in light of further research, be able to make prescriptions on the basis of an individual's genes. This will offer far greater precision. Even so, statements about the human anatomy will remain useful as a rough and ready guide. The human kind will remain natural enough to stand out as valuable for a rich variety of everyday medical explanations.

*All* biological groups, it must be conceded, are unnatural to some extent. The members of a species or other taxon have a lot in common, which allows them to figure in explanations about ecology, physiology, anatomy, behavior, developmental biology, and so on. Even so, generalizations must be qualified, because generalizations about particular species are full of exceptions, as many have observed (see, e.g., Rosenberg 1987, p. 195). Individuals fall outside the mainstream because of both environmental and genetic differences. Injury or disease may, for example, result in the loss of a limb. Ordinary genetic recombination lies behind many well-known defects and anomalies, like anencephaly, the absence of a brain at birth. And mutation may cause various changes, such as a fruit fly's growing four wings instead of two.

All of these sources of variation render even the members of a species *at a given time* a somewhat heterogeneous group. When time is taken into consideration, the deficiencies of taxa for the sake of explanation become still clearer. There will be more variation, of course. More importantly, uniqueness will be lost. The tree of life displays gradual change, but systematists break the tree into discrete units. So it is unrealistic to expect neat generalizations that apply to a taxon's members but not to outsiders, including recent ancestors.

Being a member of a taxon is not a precise necessary and sufficient condition for having many significant properties. No doubt explanations appealing to taxa would be more complete and precise if matters were otherwise.

All of this suggests that perhaps we should turn away from biology and toward physics and chemistry for natural kinds. *Mass*, *quark*, *water*, and *gold* seem decidedly less hodgepodge than biological taxa and therefore seem to be better examples of natural kinds than *tiger* and *oak*. Some theorists have followed reasoning like this to the conclusion that even chemistry is bereft of natural kinds. Paul Churchland, for example, admits at most only a few items like *mass*, *length*, *duration*, and *charge* to his list of natural kinds.[17]

But naturalness, as I have stressed, comes in degrees: Kinds can be more or less natural. I have argued that biological taxa are at least somewhat natural, comparing favorably to kinds like the named-on-a-Tuesday kind. Even if kinds that are *more* natural may be found, it seems correct to call taxa "natural," at least in many contexts. Let me explain.

## II.3. *High Standards*

Claiming that the kinds that we have been calling "natural" are not really natural is like claiming that the items we have been calling "flat" are not really flat. Flatness, like naturalness, comes in degrees. Peter Unger (1975) has argued that nothing is flat, because any allegedly flat item is bumpy by comparison with some other real or possible item. David Lewis (1983) replies that there are different standards of precision for the appropriate use of a term like 'flat'. The standards in use are set by context. In a looser context I might call the road "flat." After a bumpy jeep ride across the savannah, for example, context would permit me to say, "It is nice to be able to ride again on the road, because it is flat." In another context it may be incorrect to call the road "flat." If you try to use the road or sidewalk to support a piece of paper you are writing on, you will be correct to say, "I need something flat, like a clipboard or a desk. This road is bumpy." In another context, say that in which the reflection of mirrors is at issue, it will be appropriate to call the desk "non-flat." What Unger does in arguing for the skeptical position that nothing is flat is to raise the standard of precision. He is right, given the context he sets, to deny the flatness of any object you care to choose. But in less precise contexts it is still correct to call the desk or road "flat."

This lesson from Lewis suggests application to discussions about what kinds are natural. In stricter contexts, it seems correct to deny that biological taxa are natural, even if it is okay to call *charge* or *mass* or even $H_2O$ "natural." Biological kinds are somewhat hodgepodge by comparison. But it does not follow that philosophers are incorrect to say that species are "natural": There are less strict contexts.

Philosophers like Churchland set very high standards. For Churchland, biological taxa are not natural no matter how faithfully they represent the tree of life, because the tree of life and its particular form are cosmic accidents.

> One does not naturally think of biological species as being as arbitrary as I am here insisting, since our world presents us with only a fixed subset of the infinite number of possible species. But that subset is an accident of evolutionary history. We could have had a completely different set of species, any one of countless other sets. (Churchland 1985, p. 12)

Churchland is right that there could have been a different tree of life. So theorists interested in uncovering the basic laws underlying all actual and physically possible phenomena will have no use for biological taxa. In the context of inquiries like the foregoing, a kind can correctly be called "natural" only if it figures in the very "most basic laws of an all-embracing physics," as Churchland maintains. By that standard, what are ordinarily called "natural kinds," including biological taxa, fail to measure up. What kinds remain form, at best, "a very small, aristocratic elite," and Churchland concedes that "perhaps there are no natural kinds at all" (1985, p. 16).

Relative to the context Churchland sets, his claims are fine. But there are other contexts, set by other scientific projects. Darwin sets another context when he says that organisms "can be classed either naturally according to descent, or artificially by other characters" (Darwin 1871, vol. 1, p. 60).[18] Darwin is interested not in exposing the most basic laws of physics but rather in reconstructing the actual tree of life. When this task is at issue, it is surely acceptable to say that some biological groups are more natural than others. A division of organisms by color would yield highly unnatural groups: The group constituted just by red organisms, like red cardinals, red guppies, and red aquatic plants of the species *Rotala macranda*, is not a group with any evolutionary significance. The birds do, on the other hand, constitute a group with evolutionary significance. It would be correct to say, in the right context, that they constitute a "natural" taxon, well suited to a biologist's evolutionary explanations.

## II.4.  *Low Standards*

The recognition that contexts can be tightened and loosened does much to reconcile the impulse to concede the power of skeptical arguments that biological kinds are not natural with the commonsense impulse to say that the traditional biological kinds are natural. It also clears the way for counting as natural the kinds from various less precise or less developed sciences, as various philosophers have recommended (e.g., Boyd 1991). 'Depression', 'inflation', 'monopoly', and 'money' are good candidates from economics. Weather science has its fare ('atmospheric pressure', 'tornado'), geology has its fare ('continental plate', 'earthquake'), and so on.

What are called "natural kinds" can be less natural than these. There is a somewhat isolated tradition in the literature that associates "natural-kind terms" with ordinary, as opposed to grue-like, or "gruesome" kind terms or predicates. 'Greenkind' names a kind that has as members at any time just those objects that are green at that time. Compare this with 'gruekind' (adapted from Goodman 1955). Gruekind has as members just those objects that are green and examined before tomorrow or blue and examined after tomorrow.

Gruekind is not projectible: It does not belong in the conclusion of inductive arguments in the way that greenkind does. Even if all emeralds ever observed have belonged to gruekind, we cannot legitimately infer that future ones will. Greenkind is projectible. If all emeralds ever observed have belonged to greenkind, we can legitimately infer that future ones probably will, too.

Some authors say that greenkind is natural, because it is nongruesome and it is projectible. These philosophers would certainly expand the *usual* list of natural kinds. Greenkind is taken to be the paradigm of an *artificial* kind in most discussions of natural kinds. Greenkind's members include green trees, green frogs, green erasers, and green lakes. These hardly form a very natural extension. Members of the polar bear species have much more in common. The use of the term 'natural kind' would become too watered down to satisfy its usual purposes if 'natural kind' were to be applied in all contexts to kinds as unnatural as greenkind. Many authors have noticed this. Poncinie (1985, p. 417), for example, writes that an "interest in Goodman's paradox" has led some philosophers to count as natural kinds items that most speakers interested in natural kinds would not count, like color-kinds. "This is a different usage than the one I am taking from Kripke and Putnam," says Poncinie. Van Brakel similarly writes that "having projectible properties doesn't make something into a natural kind, as the concept is normally understood" (1992, p. 253). He points out that *red paint* is projectible. Still, it hardly counts as a natural kind in the way that a species or chemical substance does.

On the other hand, color-kinds do lend themselves to some explanations. Greenness often provides camouflage. Thus, a frog's membership or nonmembership in greenkind can explain why it survived or not in a particular habitat. Objects belonging to blackkind tend to absorb heat, which would explain why a black suit is a poor choice in hot weather. Given such explanations, I seem to be committed, by my own account of a natural kind as an explanatory kind, to watering down the use of 'natural kind' to the point that even greenkind and blackkind belong to its extension. This would be problematic.

But again, context can provide a way out of the dilemma. Greenkind is *more* natural than gruekind. Therefore in a sufficiently loose context, in which gruesome kinds serve as a foil, it could be proper to call greenkind "natural." No one ever explains why any member of a gruesome kind has this or that characteristic by explaining, "It is a member of gruekind." Nor does anyone explain anything about why the members of gruekind are in general thus and so. Gruesome kinds are not useful for giving explanations.[19] It is different for ordinary color-kinds. These do lend themselves to some explanations. Greenkind and blackkind are too unnatural to be called "natural" in the typical contexts in which *I* will be interested. Following the dominant tradition, I will employ a stricter context for the use of 'natural kind'. But there are looser contexts. Speakers from other traditions, who count greenkind as "natural," speak accurately because they set a less strict context for the use of 'natural'.[20]

In the same way, when I suggest that kinds like *toothpaste* and *trash* are not natural kinds, I speak in a rather strict context. Although *toothpaste* and *trash* do figure in some explanations, these kinds are not on a par with kinds like *polar bear* or *water*: The latter kinds allow for more explanations or more scientifically interesting explanations than do the former. Because *polar bear* and *water* allow for more explanations or more scientifically interesting explanations than do *toothpaste* and *trash*, it is correct to call *polar bear* and *water* "natural" in contexts in which such theoretical explanations are at issue, whereas it is incorrect to call *toothpaste* and *trash* "natural" in these contexts.

There are, however, looser contexts in which the naturalness of *toothpaste* and *trash* becomes salient. These kinds are more natural than the *named-on-a-Tuesday* kind or the gruesome kind *toothprash*. In contexts in which highly unnatural kinds like the latter serve as a foil, it is permissible to call *toothpaste* and *trash* "natural."

Typical philosophical discussion about "natural kinds" takes place in a fairly strict context in which kinds like *toothpaste* and *trash* do not measure up to the high standards set for the proper use of 'natural'. When *I* speak of "natural kinds" in this book, I speak in such a context. I adhere to the moderately strict standards of a well-established philosophical tradition: Those standards have been set mainly by the discussion and examples of Kripke, Putnam, and others. In the context of this tradition, *toothpaste* and *trash* are not properly called "natural kinds"; they are properly called "artificial kinds." But the standards employed in the context are not so strict as to make it improper to call *polar bear* and *water* "natural." These are properly called "natural" in the context. Indeed, they are paradigmatic "natural kinds."

## II.5.  *Natural Kind Terms from the Vernacular*

I have argued that taxa are natural, and I have suggested that the taxa recognized by different systems of classification may be natural in different respects. Given that taxa are natural kinds, there are natural-kind terms from biology. These include, for example, '*Raphanus sativus*' and '*Panthera tigris*'. But one might wonder whether they include terms from the *vernacular*, which are not really taxonomic terms. Scientists did not coin terms like 'radish', 'tiger', 'oak', and 'bear'. Lay speakers coined these terms, and they are part of lay speech, unlike scientific terms for taxa that are in Latin form.

Does the vernacular supply natural-kind terms? I will defend an affirmative answer. Biological kind terms from the vernacular have a meaning that matches or at least closely *approximates* the meaning of scientific terms. Therefore, in suitable contexts it is accurate to call such terms "natural-kind terms."

Some philosophers have cast doubts on the claim that the use of vernacular terms approximates the use of scientific terms. A couple of worries are voiced with some frequency. One worry concerns whether vernacular terms for biological kinds have extensions that are unnatural from a biological point of view. The other worry concerns whether, even if vernacular terms have extensions that are natural from a biological point of view, this is only a coincidence, because the conditions for the proper application of vernacular terms do not have much to do with scientists' empirical findings about evolution and the like, whereas the conditions for the application of scientific terms have everything to do with scientists' empirical findings about evolution. I address these concerns in order below.

II.5.a.  DO VERNACULAR TERMS HAVE NATURAL EXTENSIONS?  For the argument against my position that vernacular terms have scientifically respectable extensions, one naturally turns to John Dupré (1993, pp. 26ff.). Dupré offers many examples to illustrate that there is often no correspondence between science and the vernacular. 'Onion', he says, does not have an extension that is natural from a biological point of view, nor does 'garlic'. The distinction between onions and garlic is not recognized in scientific taxonomy: An important culinary distinction is ignored by science. Onions also play an important role in Dupré's case for the artificiality of another example that has been cited frequently: 'lily'. Dupré says that 'lily' cannot refer to the botanical family Liliaceae because that family includes onions, which are not lilies. Something similar is said about some terms for animals, like 'hawk', which is supposed to be unnatural because otherwise vultures would belong to its extension.

I do not think that any argument along these lines can show that the vernacular lacks biological natural-kind terms. First, Dupré's litany of terms for kinds that are allegedly unnatural from a biological point of view seems vulnerable to criticism. Second, even if his litany stands, it is more or less irrelevant to the naturalness of many other kind terms. I will address these criticisms in reverse order.

*II.5.a.i. Natural Kind Words Outside the Litany.* No one has ever suggested that all or even the great majority of vernacular terms for biological kinds name natural kinds: Consider 'shrub', 'tree', 'weed', 'livestock', and 'vermin'. A term coined to refer to a natural kind but that has turned out not to do so, at least by purely genealogical standards, is 'reptile'. I discuss other similar terms in Chapter 3. Everyone acknowledges that vernacular terms for biological kinds may sometimes fail to name natural kinds, either by design or by accident. Therefore if terms without scientific correlates, like 'lily' perhaps, are to present a problem for well-known accounts of vernacular natural-kind terms, that will have to be because there are so many of them. If hardly any biological kind terms from the vernacular turn out to have natural extensions, then generally philosophical accounts of vernacular natural-kind terms will fail to have much application to biological terms. That will be a problem for those accounts.

But it is not the case that few biological kind terms from the vernacular have natural extensions. A great many vernacular terms for biological kinds have highly natural extensions. Typically, a vernacular term with a natural extension shares its extension with a corresponding scientific term. I have already suggested as much for 'radish', the corresponding scientific term for which is '*Raphanus sativus*', and for 'tiger', the corresponding scientific term for which is '*Panthera tigris*'. 'Oak' and 'bear', also mentioned a few paragraphs back, have correlates from science as well, the respective scientific terms being '*Quercus*' and 'Ursidae'.[21] Other terms with natural extensions are readily available, because ordinary speakers of the vernacular have many practical reasons for naming various natural kinds. A sample of vernacular terms for animals might include 'giraffe', 'polar bear', 'bullfrog', 'monarch butterfly', and 'sockeye salmon', all of which refer to species; and 'ant', 'spider', 'bird', 'snake', and 'animal' itself, all of which refer to higher taxa. Terms for plant species might include 'carrot', 'spinach', 'blueberry', 'poison oak', and 'sugar maple', whereas terms for higher taxa from the plants might include 'cactus', 'apple', 'sunflower', and 'plant' itself. There are also some terms for organisms that are neither plants nor animals, like 'fungi' and '*E. coli*'. Both 'fungi' and '*E. coli*', as well as many other names, such as '*Boa constrictor*' and '*Tyrannosaurus*', illustrate that sometimes scientific terms just are vernacular, rather than there being a separate name from the vernacular corresponding to a scientific name.

The foregoing terms and many more like them have extensions that include just the members of biological taxa. Although one might expect some terms to turn out to have unnatural extensions, as Dupré's examples are supposed to have done, that sort of outcome is nowhere close to being the rule: A great many purported natural-kind terms have turned out to have the expected natural extensions.

The interests of lay speakers tend to influence lay speakers' decisions about what natural kinds to name, as the foregoing examples indicate. Speakers are more apt to name species of large mammals or crop plants than arthropods. But a kind can be perfectly natural from a biological perspective even when the decision to *name* it, rather than to leave it unnamed, is made in part on the basis of nonbiological reasons. Thus, although lay interests have influenced speakers to coin terms like 'tiger' and 'radish' rather than other terms for obscure arthropod species, the foregoing terms still seem to refer to natural kinds. It may surprise some readers to learn that scientists themselves are sometimes moved by factors like economic importance when they decide which of various evolutionary groups should be assigned taxonomic names (de Queiroz and Cantino 2001, p. 262). But the groups that scientists name are natural from a biological perspective even when economic considerations have played a role in the decision to name them rather than to name other currently unnamed natural groups instead.

*II.5.a.ii. The Litany of Artificial Kinds.* Plausible examples of natural-kind terms are found even within Dupré's litany of allegedly artificial-kind terms, though these tend to present more delicate cases. The claim, repeated by many authors after Dupré, that onions and garlic are not distinguished by scientists is inaccurate. Although onions and garlic belong to the same genus, they are different species. 'Onion' refers to *Allium cepa* and 'garlic' to *Allium sativum*.[22]

It is true that sometimes other plants are called "onion" or "garlic" with a qualifying word. For example, *Allium vineale*, a wild weed taken seriously by agriculture specialists, is called "crow garlic." This raises a common complication in the study of vernacular terms. Often a term is used differently in different places, takes on a different meaning in compound expressions (compare 'hot dog' or 'bush baby', for example), and so on. But such complications do not threaten the terms' reference to natural kinds in various specific contexts. A more familiar example might help to make the point clearer. When speakers use the term 'lion' by itself, they almost always use it to refer to the great African cat *Panthera leo*. Thus, when speakers report having seen photographs of lions on the African savannah, or when they talk about the great manes of lions, or about lion prides, they are talking about *Panthera leo*. Still, speakers often call the cougar (*Felis concolor*), a smaller, solitary American cat that lacks a mane, a "mountain lion," and speakers even call insects of the family Myrmeleontidae "ant lions." It is clear that the group including just *Panthera leo* and *Felis concolor* is not a natural kind, let alone the group that includes these two cats plus Myrmeleontidae. But this fact cannot be used to mount much of an objection against the claim that 'lion' refers to a natural kind, because it does typically refer to the natural kind *Panthera leo* when used outside of compound expressions like 'mountain lion' and 'ant lion'.[23] Similar considerations seem to apply to 'garlic', which, when used alone by cooks and ordinary speakers, refers to *Allium sativum* despite the fact that speakers in agriculture use 'crow garlic' for the weed *Allium vineale*. Other names for *Allium vineale* include 'wild garlic' and 'false garlic', the latter of which certainly suggests that this is not the authentic kind usually called "garlic."

I do not, of course, suggest that context always clears up vagueness. Sometimes context will leave matters vague, too.[24] It is just that often context does rule out interpretations according to which a term has an unnatural extension. The vernacular terms 'onion' and 'garlic' have natural extensions as they are used in ordinary contexts.

Much the same can be said of other examples Dupré adduces, including even the well-known example 'lily'. The scientific correlate of 'lily' is not obvious: Candidates include the order Liliales, the family Liliaceae, and the genus *Lilium*. Some so-called "lilies," like the water lily, belong to none of these nested groups, being unrelated to true lilies on anyone's account. But which of the foregoing groups comprises what are properly called "the true lilies"? This seems to depend on the context. Most authorities restrict 'lily' to the genus *Lilium*, to which onions, and for that matter tulips and asparagus, do *not* belong. Onions, tulips, and asparagus have been placed in Liliaceae and Liliales but not in the narrower *Lilium*. Thus, in typical contexts 'lily' has an extension that does *not* contain onions but that *is* nevertheless *natural*, rather than gerrymandered to exclude the onions. That extension includes just the plants in *Lilium*.[25]

Although the members of the genus *Lilium* are most often counted as the true lilies, 'lily' is sometimes used for the entire lily family or order. When the family or order is at issue, it seems right to say that onions, tulips, and asparagus belong to the lilies, though 'lily' is vernacular. Such broad use is not as common as a more restrictive use of 'lily', but people who know something about lilies or onions, tulips, or asparagus do sometimes use the term 'lily' in this broader way.[26]

'Lily' seems to have a natural extension, though one that changes with the context of use. A similar treatment suggests itself for other examples on Dupré's list, like 'hawk'. This type of example is subtle. But, as I have emphasized previously (in §II.5.a.i), there are also many relatively straight-forward cases of vernacular terms from English with natural extensions in the biological world. It seems wrong to deny that the vernacular is a source of natural-kind terms on the grounds that the extensions of vernacular terms are unnatural.

II.5.b.  THE RELEVANCE OF SCIENCE TO ORDINARY CLASSIFICATION. I have argued that many vernacular terms for biological kinds have natural exten-sions. Still, could this be fortuitous? One commonly articulated worry about whether there are natural-kind terms in the vernacular is that whereas scientists classify according to scientific criteria, ordinary speakers classify according to nonscientific criteria. Various philosophers have claimed that evolutionary findings in particular are irrelevant to the use of vernacular terms, though they are obviously highly relevant to the proper use of scientific terms.[27] If that is true, then although 'bear' and 'Ursidae' have a common extension that comprises the actual bears in the world, which form an evolutionary group, it is only by chance that 'bear' has a scientifically interesting extension. If on Alpha Centauri we were ever to find an animal that closely resembles our terrestrial bear, this line goes, the animal would *not* belong to the family Ursidae, but it *would* be a *bear*. So-called "natural kinds" from the vernacular have little in principle to do with the kinds recognized by scientists.

I cannot accept such a position. The connection between scientific findings and vernacular categorization is substantially tighter than this. There is, as Putnam observes, a division of linguistic labor. Lay speakers defer to experts' judgment about what belongs in the extension of a term like 'bear'. Scien-tific influence on lay speech is commonplace. Ordinary speakers are happy, for example, to grant that whales are not fish. No extraterrestrial bear-like organisms have turned up to test speakers' dispositions, but explorations in Australia did uncover the koala, which looks like a bear. Although many lay speakers do call this creature a "koala bear," they seem prepared to con-cede that it is not *really* a bear. At least they concede that it is not really a bear when they become informed of the koala's scientific status. Vernacular speakers educate one another about that status. Examples of such instruction are ubiquitous. You may find trivia about "The bear that's not" even on the packaging of your instant oatmeal, as I have. My illustrated package went on to say, "Did you know the koala bear isn't a bear at all? The cuddly creature is really a marsupial – an animal that carries its young in a pouch" (Quaker Oats Company 1997). The literature of a stuffed-animal toy store similarly says, "Koalas aren't bears. They aren't even related to bears." The source goes on to say that "The reason that the koala is called a koala bear is because the koala looks like a teddy bear."[28] These publications are hardly addressed to scientists.

Instruction on the proper use of 'bear' like the foregoing abounds in vernacular speech and publications. In light of vernacular speakers' own concessions about whether or not koalas are real bears, it would seem best to say that they are not. It would seem best to say that 'koala bear' is a misleading expression, something like 'sea horse', 'ant lion', or 'bush baby'. Some readers may suspect that there is a looser sense, or anyway a different sense, of 'bear', 'horse', 'lion', and 'baby', on which the extensions of these words really do include koala bears, sea horses, ant lions, and bush babies, respectively. This seems questionable (see note 23), but even if it is true, there would be stricter or more proper vernacular senses, according to which 'bear' fails to designate marsupials, 'horse' fails to designate sea horses, and so on. These senses would vindicate the status of 'bear', 'horse', and so on as natural-kind terms.

Enough about koalas. As a point of contrast, consider the giant panda. Like the koala, the panda was discovered after the term 'bear' had long been in use: The panda was first described in 1869 by a French missionary to China. His description sparked a debate among scientists that would continue for more than a hundred years: "Is the panda a bear? Is it a raccoon? Or does it actually belong to a family of its own?" (O'Brien 1987, p. 102). Genetic studies have confirmed that the giant panda shares an evolutionary history with the bears, so it is counted as a bear. "The Giant Panda Is a Bear," as the title of one article addressed to zoo enthusiasts confirms. It is true that "you shouldn't call the koala a bear (it's a marsupial), but you *can* call the giant panda a bear!" (Ryder 1987, p. 16). In light of the tendency of experts to instruct the lay public about the proper use of vernacular terms like this, and the tendency of lay speakers to accept this sort of instruction, it seems evident that there is a division of linguistic labor. It is highly doubtful that evolution is just irrelevant to an ordinary speaker's use of 'bear'. Lay terms do reflect scientific categorization. They approximate scientific terms in their reference. They are appropriately called "natural-kind terms."

# 2

# Natural Kinds, Rigidity, and Essence

In Chapter 1, I argued that 'water', 'horse', and 'mammal' are natural-kind terms. Along the way I argued briefly that biological kinds have theoretically interesting essential properties, or properties that they possess necessarily and not just contingently. It is time for a more thorough look at necessity in connection with kinds. That is the task of the present chapter.

Although I will be taking a close look at essences, I will postpone a serious look at whether scientists' conclusions about essences are empirical *discoveries*, as they are commonly thought to be. A serious treatment of that issue, and surrounding issues about conceptual change in science and philosophy, will have to wait for the chapters following this one. For now, I will address just the issue of whether interesting statements about natural kinds are indeed necessarily true, not whether they have been discovered to be so.

My answer, as I have already indicated, will be that at least some interesting statements about natural kinds do indeed seem to be necessarily true. In section (I), I discuss identity statements. I clarify and defend the position that natural-kind terms are rigid designators, so that some statements of identity about natural kinds are necessarily true. In section (II), I discuss whether individuals essentially belong to the kinds to which they belong and whether narrower kinds essentially belong to the broader kinds to which they belong – for example, whether Tabby the tiger essentially belongs to *Panthera tigris*, and whether the species *Panthera tigris* itself essentially belongs to the broader kinds Mammalia, Chordata, and so on. I follow these sections with a brief conclusion.

## I. RIGIDITY, IDENTITY, AND NECESSITY

Saul Kripke has convinced the English-speaking philosophical community that some statements of identity are necessarily true but knowable only a posteriori. There are a few dissenters, but by and large philosophers agree that 'Hesperus = Phosphorus' is necessarily true, even though it took empirical investigation on the part of astronomers to determine that statement's truth. 'Cicero = Tully' is another example of a necessarily true identity statement, although this statement was presumably knowable a priori for some speakers.

In this section, I go over briefly the familiar Kripkean arguments for the necessity of statements like 'Hesperus = Phosphorus', which are about individuals, and then I proceed to clarify how these arguments, which concern the notion of rigid designation, apply to statements containing names for *kinds*. Kripke first sketched some outlines of his ideas about kind identities more than a quarter of a century ago, but after first sketching his ideas he has never revisited them in order to supply needed development or to defend the ideas against the many criticisms that they have generated. He has left a lot of work to be done. I will develop and defend a generally Kripkean line with respect to identity and necessity concerning kinds, beginning with simpler cases of identity, and building up to more difficult ones, responding to objections as I proceed.


I.1. *On the Rigidity of 'Hesperus' and 'Phosphorus'*

As I have said, Kripke's arguments for the necessity of 'Hesperus = Phosphorus' have convinced the overwhelming majority of philosophers. I review those arguments here (in §I.1), so that afterward (in §I.2–§I.9) I can discuss their application to statements containing kind terms, about which there is much more doubt and controversy.

The arguments for the necessity, in case of truth, of 'Hesperus = Phosphorus' are as follows (Kripke 1980, pp. 100–5). 'Hesperus' and 'Phosphorus' are both names for the planet Venus. The ancients named Venus twice, not realizing that they had done so. They associated 'Hesperus' with the body when they observed it in the evening, and 'Phosphorus' with the same body when they observed it at a different location in the morning, supposing this to be a different body. So they would have agreed to sentences like 'Hesperus is the object over yonder' and 'Phosphorus never appears in the sky when Hesperus does'. They would not have agreed with the sentence 'Hesperus = Phosphorus'.

Nevertheless, astronomers later came to realize that 'Hesperus' and 'Phosphorus' name the same body, so that Hesperus is Phosphorus. This was an empirical discovery. It was an empirical discovery of a necessary truth. It is not necessary that 'Hesperus' and 'Phosphorus' should name the same object, of course. But given our use of the terms, 'Hesperus = Phosphorus' is necessarily true if true at all.

Why is this sentence necessarily true if true at all? After all, not all true identity statements are necessarily true. Consider the sentence

H. 'Hesperus = the brightest nonlunar object in the evening sky.'

This sentence is true. At least it is true if evening is understood to start after sunset, so let evening be so understood. The sentence is true, but it is only contingently true, because Hesperus might have been overshadowed by some other still brighter object, had our solar system had more planets, including one brighter than Hesperus. Or, some other *actual* planet, Neptune, perhaps, might have outshone Hesperus had Hesperus been obscured by cosmic dust. So the above identity statement (H) might have been false.

Kripke's well-known response to this difficulty is that even though (H) is contingent, 'Hesperus = Phosphorus' is necessarily true, because 'Hesperus' and 'Phosphorus' are both names for an object, and thus *rigid designators*: Roughly, that is to say that each designates the same object in every possible world, or at least in every world containing the object.[1] Thus, 'Hesperus' is different from the description 'the brightest nonlunar object in the evening sky'; this description designates Hesperus, but not rigidly, because it does not designate Hesperus in all possible worlds. There are possible worlds in which *Neptune*, and not Hesperus, satisfies that description. Again, Neptune might have been the brightest such object in the evening sky, which is to say that in some possible worlds, it is the brightest. In those possible worlds, 'the brightest nonlunar object in the evening sky' designates Neptune. In still other possible worlds, 'the brightest nonlunar object in the evening sky' designates Pluto. 'The brightest nonlunar object in the evening sky' designates different objects in different possible worlds: Hesperus in the actual world, Neptune and Pluto in others.

'Hesperus', on the other hand, does not designate different objects in different possible worlds. It does not designate Neptune in some worlds and Pluto in others. In describing worlds in which Neptune is the brightest celestial body, we apply 'the brightest nonlunar object in the evening sky' to Neptune rather than to Hesperus, but we still apply 'Hesperus' to Hesperus, not to Neptune. In describing different counterfactual situations we use 'Hesperus' for the same object, Hesperus, though we use 'the brightest nonlunar object in the evening sky' for different objects. 'Hesperus' names Hesperus in all possible worlds. It is rigid.

Now, because a rigid designator picks out the same object in all possible worlds, an identity statement in which the identity sign is flanked by two rigid designators must be necessarily true if it is true at all. For if, say, the identity statement 'Hesperus = Phosphorus' is true, then 'Hesperus' and 'Phosphorus' name the same object. As it happens, that object is Venus. Moreover, because 'Hesperus' and 'Phosphorus' are rigid, each names just the object it *actually* names in all possible worlds in which it names anything. Each names Venus in all possible worlds. Because 'Hesperus' and 'Phosphorus' both name Venus in all possible worlds, and because Venus = Venus in all possible worlds, 'Hesperus = Phosphorus' is true in all possible worlds.

## I.2.  *Two Types of Identity Statements for Kinds*

The familiar foregoing argument for the necessity of identity statements involving proper names for individuals is supposed to extend to theoretical identity statements like 'Gold = the element with atomic number 79' and 'Water = $H_2O$' (Kripke 1980, pp. 138–40, 148). Kripke does not give a biological example of a theoretical identity statement, but I suggest, in light of my previous discussion (Chapter 1, §I.3), that plausible candidates would be 'Mammalia = the clade that stems from the ancestral group $G$' and 'Aves = the clade that stems from the ancestral group $A$'. $G$ and $A$ are ancestral groups that have given rise to all other mammals and birds, respectively.[2]

Theoretical identity statements are supposed to be necessary if true at all, as a result of containing two rigid designators. Thus Kripke: "Theoretical identities, according to the conception I advocate, are generally identities involving two rigid designators and therefore are examples of the necessary *a posteriori*" (1980, p. 140).[3] But, as I have suggested, whether the arguments for the necessity of statements about concrete individuals apply to statements about kinds has often been questioned. Many problems, criticisms, and complications have been raised in the literature. Some of these troubles could have been circumvented had Kripke focused upon more modest identity statements.

I.2.a.  DESCRIPTIONS VS. NAMES FOR KINDS.  Some concerns are invited by the striking dissimilarity between sentences like 'Hesperus = Phosphorus' and 'Cicero = Tully' on the one hand, and sentences like 'Gold = the element with atomic number 79' or 'Mammalia = the clade that stems from the ancestral group $G$', on the other. 'Hesperus' and 'Phosphorus' are ordinary names for an object, as are 'Cicero' and 'Tully'. 'Gold' and 'Mammalia' are ordinary names for kinds, but 'the element with atomic number 79' and 'the clade that stems from the ancestral group $G$' do not seem to be names. They seem to be descriptions.

When applying his lesson about individuals to kinds, Kripke could have forestalled certain complications and doubts by discussing statements more straightforwardly analogous to 'Hesperus = Phosphorus' than the sentences about kinds on which he expounds. A statement much more closely analogous to 'Hesperus = Phosphorus' would be '*Brontosaurus = Apatosaurus*', because '*Brontosaurus*' and '*Apatosaurus*' are two *names* for the same kind of dinosaur, just as 'Hesperus' and 'Phosphorus' are two names for a planet. It is not clear why Kripke discusses only *theoretical* identity statements for kinds. These theoretical identities resemble 'Cicero is a product of the egg and sperm that generated him' rather than 'Cicero = Tully' in the respect that they aim to expose essence. Kripke defends the necessity of both 'Cicero is a product of the egg and sperm that generated him' and 'Cicero = Tully' (1980, pp. 110ff.), but his defense of the necessity of 'Cicero = Tully' is much less disputed. In the same way, the claim that '*Brontosaurus = Apatosaurus*' is necessarily true invites less dispute than the claim that '*Brontosaurus* = the clade with such and such stem' is necessarily true.

In the next few sections (§I.2–§I.5), I concentrate on statements like '*Brontosaurus = Apatosaurus*'. Only afterward do I address theoretical identity statements (in §I.6–§I.9). In the remaining paragraphs of the present section (I.2), I suggest that '*Brontosaurus = Apatosaurus*' is a biological version of 'Hesperus = Phosphorus'. Both statements seem to have been discovered by empirical investigation to be necessarily true. And the necessity of each statement seems to obtain in virtue of its containing two rigid designators. I clarify and defend the claim that '*Brontosaurus = Apatosaurus*' contains two rigid designators in sections (I.3)–(I.5).

I.2.b.   A PARALLEL TO 'HESPERUS = PHOSPHORUS'. The statement '*Brontosaurus = Apatosaurus*' has a history quite like that of 'Hesperus = Phosphorus'. '*Brontosaurus*' was coined as a genus-term in 1874 by O. C. Marsh, who thought he had discovered a new genus of dinosaur in Wyoming. As it turns out, the fossils he discovered were fossils of a dinosaur genus that he himself had already discovered and named '*Apatosaurus*'. Marsh supposed that the specimen he associated with the name '*Brontosaurus*' and the specimen he associated with the name '*Apatosaurus*' could not be from the same genus because there was such a difference in size between the two specimens. He did not realize that the reason for the difference in size was only that one of his specimens was not fully grown. Later, another scientist, Elmer Riggs, straightened out the matter, determining that Marsh had applied two names to the one genus. Riggs discovered empirically that '*Brontosaurus = Apatosaurus*' is true.

Because '*Brontosaurus* = *Apatosaurus*' is true, it must be necessarily true, assuming that '*Brontosaurus*' and '*Apatosaurus*' are rigid designators, as they seem to be. This sentence seems to be a biological version of 'Hesperus = Phosphorus': Both 'Hesperus = Phosphorus' and '*Brontosaurus* = *Apatosaurus*' were discovered, by empirical means, to be necessarily true.

As I have suggested, the lack of any discussion about this kind of case in the philosophical literature to date indicates a significant and surprising oversight. The history of biological taxonomy supplies many plausible examples of inadvertent multiple namings just like that for 'Hesperus' and 'Phosphorus', but for kinds, and these generate less-problematic instances of necessarily true statements about natural kinds than the usual theoretical identities, as I have suggested. They also generate far more likely instances of *empirically discovered* necessary truth.[4]

## I.3.    *What Do Kind Terms Rigidly Designate?*

I have said that '*Brontosaurus*' and '*Apatosaurus*' appear to be rigid designators. 'Whale', 'water', and 'gold' all appear rigid as well. But I have yet to substantiate my position that these terms are indeed rigid. I turn to that now. A term like '*Brontosaurus*' or 'water' is rigid if and only if it designates the same entity in all possible worlds. But what entity could a kind term designate in all possible worlds? Certainly not its extension, which is the set of individuals to which it applies. The extension of 'whale' is the set of actual whales out there in the ocean, including Gigi and Flipper. But there are worlds in which Gigi and Flipper never come into existence, but other whales, which have not in fact come into existence, do. 'Whale' has a different extension in such worlds. Similarly, in the case of water, there are possible worlds in which the water that actually exists does not exist: 'Water' has an extension in (some of) those worlds, however. Its extension in those worlds consists of water particles that as a matter of fact do not exist, but might have. And the same applies to 'gold', '*Brontosaurus*', and so on.

A kind term does not rigidly designate its extension; that varies from world to world. What it rigidly designates, I propose, is a *kind*. '*Brontosaurus*' designates the same *abstract kind* in every possible world, the brontosaur kind, even though the concrete individuals instantiating the kind, and constituting the term's extension, vary from world to world.[5]

If, as I propose, kind terms rigidly designate kinds, then, contrary to what would naturally be expected, natural-kind terms are not *unique* among kind terms in their rigidity: Even terms for *artificial* kinds are rigid. Just as each of the terms '*Brontosaurus*' and 'gold' designates the same abstract kind in all possible worlds, despite the fact that the extension of each term varies from world to world, so each of the terms 'lawyer', 'bachelor', and 'soda pop' also designates the same respective kind in all possible worlds, despite the fact that extensions vary from world to world. 'Lawyer' designates the lawyer kind in every possible world. The extension of 'lawyer' differs from one world to the next: Johnnie Cochran is a member of the extension of 'lawyer' as things are, but he might have become a mail carrier instead, so there are worlds in which 'lawyer' has a different extension. Still, had Johnnie Cochran become a mail carrier as his sole occupation, he would not have belonged to the extension of 'lawyer', because 'lawyer' designates the lawyer kind in all possible worlds.

On my proposal, kind terms are rigid if and only if they designate the same kind in all possible worlds, so artificial-kind terms are rigid, because artificial-kind terms do that. Some philosophers think that this dooms the proposal. The extension of rigidity to artificial-kind terms is startling, I grant, but it does not amount to the "glaring problem" to which it is often supposed to amount (Schwartz 1980, p. 190; see also, e.g., his 1977, pp. 37ff.). The common fear is that a generous interpretation of rigidity like mine, according to which even artificial-kind terms are characterized as rigid, renders rigidity useless. According to the interpretation that I endorse, the worry goes, rigidity would be incapable of performing the work that rigidity is supposed to do. This worry must be addressed.

### I.4.  *Rigidity without Power?*

One reason it might be supposed that artificial-kind terms cannot be counted rigid is that to count them rigid would trivialize rigidity by counting *every* kind designator rigid. Surely some designators for kinds are intended to be nonrigid, or else there would apparently be no reason to take special note of the rigidity of certain kind designators, as if to distinguish them. Because, according to this worry, my preferred account of rigidity makes every kind designator come out rigid, something appears to be wrong.

Something like this criticism seems to be what worries a number of authors about proposals like the foregoing. Schwartz asserts that "Just as with singular terms there is a contrast between proper names and definite descriptions, in that the former are rigid whereas the latter are non-rigid, with general nouns there is a contrast between those that are natural kind terms and those that are not" (1980, p. 196). He thus objects to proposals that extend rigidity to unnatural-kind terms. Danielle Macbeth seems to have something like this concern in mind also, when she criticizes accounts like the foregoing on the grounds that they render rigidity trivial: "[I]n this trivial sense, every predicate, whether or not it is a natural kind term, is rigid" (Macbeth 1995, p. 263).[6]

This criticism fails to damage the account of rigidity that I endorse. My preferred account does not trivialize rigidity. It is simply not the case that every kind designator rigidly designates its kind on my preferred account. *'Brontosaurus'*, which names a dinosaur, seems to pick out the brontosaur kind rigidly. But other kind designators pick out this kind by way of the kind's fitting a particular accidental description. 'The largest dinosaur on a 1989 U.S. postage stamp' is an example: This expression designates *Brontosaurus*, because *Brontosaurus* happens to have been the largest dinosaur on a 1989 U.S. postage stamp. But it is not a designator that picks out in every *other* possible world the same kind that it picks out in the *actual* world. The U.S. Postal Service might have had only two dinosaurs on its stamps of 1989, rather than three; without *Brontosaurus*, there would have been only *Tyrannosaurus* and *Stegosaurus*. Had that been so, 'The largest dinosaur on a 1989 U.S. postage stamp' would have designated *Tyrannosaurus*, not *Brontosaurus*. 'The largest dinosaur on a 1989 U.S. postage stamp' happens to designate *Brontosaurus*, but it does not *rigidly* designate *Brontosaurus*.[7]

The rigidity of *'Brontosaurus'* would seem, then, to find a contrast in certain *descriptions*, just as the rigidity of names like 'Hesperus' and 'Ben Franklin' do. It would not seem to be a "trivial" matter at all that expressions like *'Brontosaurus'* are rigid; on the contrary, it is significant that such expressions are rigid, and that others, such as 'the largest dinosaur on a 1989 U.S. postage stamp', are not. It is significant for the very reason that the differences between expressions like 'Hesperus' and 'the brightest nonlunar object in the evening sky' are significant.

*'Brontosaurus'* and *'Apatosaurus'* both rigidly designate the same specific abstract entity: Each designates that entity in every possible world. So there is no world for which (using *our* language) *'Brontosaurus = Apatosaurus'* can possibly be false; in every world, *'Brontosaurus'* and *'Apatosaurus'* designate the same thing: *Brontosaurus*. Now, clearly no parallel argument can support the necessity of a statement like *'Brontosaurus =* the largest dinosaur on a 1989 U.S. postage stamp', any more than a parallel argument could support the necessity of 'Hesperus = the brightest nonlunar object in the evening sky', or, to use Kripke's favorite example of a contingent identity statement, 'Ben Franklin = the inventor of bifocals'. This is fortunate, because, just as the familiar latter statements involving individuals are true but only contingently so, the statement *'Brontosaurus =* the largest dinosaur on a 1989 U.S. postage stamp' is also true but contingently so. With respect to those worlds in which *Tyrannosaurus* is the largest dinosaur on a 1989 U.S. postage stamp, the

sentence '*Brontosaurus* = the largest dinosaur on a 1989 U.S. postage stamp' comes out false: For in those worlds, the largest dinosaur on a 1989 U.S. postage stamp is *Tyrannosaurus*, and obviously *Brontosaurus* is not identical to *that*.

The reason no argument parallel to that supporting the necessity of '*Brontosaurus* = *Apatosaurus*' can support the necessity of a statement like '*Brontosaurus* = the largest dinosaur on a 1989 U.S. postage stamp' is that, as we have seen, the second designator in the latter statement does not pick out *Brontosaurus* in every possible world but rather picks it out in some worlds and not in others, so it is not rigid. Because one of the designators in the identity statement does not rigidly designate its kind, there is no reason to take the statement to be necessary, even if it is true.

Having drawn a parallel between '*Brontosaurus*' and 'Ben Franklin', on the one hand, and 'the largest dinosaur on a 1989 U.S. postage stamp' and 'the inventor of bifocals' on the other, we are in a better position to see whether artificial-kind terms allow such a contrast. It will be noticed that the rigid/nonrigid distinction found above in natural-kind designators seems readily applicable to designators of artificial kinds, as well. The bachelor kind is apparently *rigidly* picked out by 'bachelor' but *non*rigidly picked out by the expression 'the kind most commonly broached in discussions about analyticity'. The bachelor kind, or bachelorhood, could not have failed to be the bachelor kind, or bachelorhood. But the bachelor kind could have failed to be the kind most commonly broached in discussions about analyticity. Philosophers might have preferred to appeal to 'spinster' and spinsterhood in illustrating (or trying to illustrate) analyticity. If philosophers had done so, 'the kind most commonly broached in discussions about analyticity' would *not* have designated the bachelor kind, though it happens to do so as things are. There are worlds, then, with respect to which 'the kind most commonly broached in discussions about analyticity' designates the bachelor kind, and others, such as the spinsterhood-preferring world, in which the expression designates some other kind. Thus, that expression does not rigidly designate the bachelor kind. 'Bachelor', by contrast, does. Similarly, 'soda pop', 'soda', and 'pop' all rigidly designate the soda pop kind; but 'the beverage Uncle Bill requested at our Super Bowl party' only contingently designates the kind.

The connection to necessity with respect to artificial kinds is straightforward. 'Soda = the beverage Uncle Bill requested at our Super Bowl party' is true but not necessarily true, because the second designator is not rigid; it refers to milk, or juice in some worlds, and in those worlds, the sentence is false. 'Soda = pop' is, on the other hand, necessarily true, because it is true and both designators rigidly designate soda pop.

41

I cannot agree with various suggestions, then, that artificial-kind terms stand in contrast to natural-kind terms over rigidity, just as descriptions like 'the inventor of bifocals' contrast as nonrigid designators with names, such as 'Ben Franklin'.[8] The proper contrast over rigidity is that between nonrigid descriptions for kinds, either natural or artificial, on the one hand, and rigid descriptions and names for kinds, either natural or artificial, on the other.

### I.5.  *Rigidity and Externalism*

It seems clear, then, that to extend the honor of rigidity to artificial-kind terms like 'soda' and 'bachelor' does not render rigidity trivial: Rigidity still does the work it is supposed to do in assuring the necessity, in case of truth, of identity statements containing two rigid designators. But perhaps it fails to do other important work. There is one other job that rigidity is often thought to do, and that it cannot do on my proposed account: Rigidity is often thought to be what does the work of putting meanings "out of the head," to use Putnam's phrase.

Consider a term like 'whale'. What makes a thing belong to the extension of 'whale' is not its possession of properties like *having a fishlike appearance* or *exhaling a large, visible spout of vapor* that a speaker associates with whales but rather its possession of *underlying* properties and relations that guarantee sameness of kind to paradigm whales. Just what 'whale' refers to is knowable only by empirical investigation of sample whales. 'Bachelor', by contrast, does not refer to something just in case it possesses some underlying properties possessed by sample bachelors like the Supreme Court Justice David Souter. The reference of 'bachelor' is determined, rather, by analytically associated conditions that have been established a priori to mark the bachelor status: unmarried, eligible malehood.[9] Rigidity is supposed to be what makes 'whale' different in this way from terms like 'bachelor'.[10]

But there is no need to appeal to rigidity to distinguish terms like 'whale' from terms like 'bachelor'. 'Whale' was given a causal baptism rather than an old-fashioned analytic definition. 'Bachelor' was not. *That* is the difference between terms like 'whale' and terms like 'bachelor'. This difference holds even though both terms are rigid because both refer to the same kind across all possible worlds. A causal baptism itself, and not the further possession of rigidity, is what is responsible for the reference of 'whale' to whatever possesses an underlying essence instead of to whatever satisfies descriptions like 'has a fishlike appearance' or 'exhales a large, visible spout of vapor'. It seems to be a running together of different elements of the Kripke-Putnam theory of reference, in particular the running together of rigidity with causal grounding, that has led to the belief that rigidity is what distinguishes terms like 'whale' from terms like 'bachelor'.

Putnam himself confuses the elements of the Kripke-Putnam theory of reference. Putnam calls causally grounded terms "indexical," because they designate whatever has the underlying essence of samples *around the speaker*. 'Water' and 'whale' are supposed to be indexical; 'hunter' and 'bachelor' are not, because they have analytic definitions. According to Putnam, "Kripke's doctrine that natural-kind words are rigid designators and our doctrine that they are indexical are but two ways of making the same point" (1975e, p. 234). Putnam is wrong: These are *not* two ways of making the same point. Rigidity has nothing to do with how a term gets hooked up to its referent, whether this is by way of samples around the speaker, as with causally grounded, or "indexical," terms like 'whale', or by description, as with terms like 'bachelor'. Rather, rigidity has to do with whether the term refers to the same entity in all possible worlds. "Non-indexical" kind terms, such as 'hunter' and 'bachelor', can refer to the same thing in all possible worlds just as well as "indexical" ones, like 'water' and 'whale'.[11]

I conclude that the objections to counting artificial-kind terms as rigid are not successful. My account of kind terms' rigidity, which counts artificial-kind terms rigid, survives the foregoing criticisms. It seems right to apply rigidity to certain natural-kind designators and certain artificial-kind designators, but not to other natural-kind designators and other artificial-kind designators.

### I.6.  *Beyond Names: Theoretical Identity Statements*

I have argued that '*Brontosaurus = Apatosaurus*' is necessarily true. It was discovered to be necessarily true. It is time to turn to *theoretical identity statements*, like '*Brontosaurus* = the clade with such and such an origin'. Theoretical identity statements are supposed to state the *essence* of *Brontosaurus* or Mammalia, water or gold. They do not simply contain two names for the kind.

I do not think that any of the theoretical identity statements discussed earlier in the chapter, or indeed any discussed in the literature, have been discovered to be necessarily true. That would include 'Mammalia = the clade that stems from the ancestral group *G*', 'Aves = the clade that stems from the ancestral group *A*', 'Gold = the element with atomic number 79', and 'Water = $H_2O$'. 'Gold = the element with atomic number 79' and 'Water = $H_2O$' do not even seem to me to be *true*, let alone *discovered* to be *necessarily* true. They seem to me to be false, for reasons that I provide later (in Chapter 4).

**Figure 2.1.** Convergent evolution between a nectar-feeding moth (left) and a nectar-feeding hummingbird (right). The similarity between hummingbird hawk moths and hummingbirds, in flight or even in the hand, is so great that the naturalist Henry Bates (1910, p. 102) could not convince locals that they belong to different species, and he himself often shot moths when attempting to collect birds. But the line of hummingbird hawk moths is not a line of birds and cannot become a line of birds no matter how birdlike the line becomes in appearance or even genetic structure. Systematists today classify by history and would refuse to count any animal as a bird unless it shares a history with the birds. Convergence is discounted. From Leakey (1979, p. 207).

'Mammalia = the clade that stems from the ancestral group $G$' and 'Aves = the clade that stems from the ancestral group $A$' are different. These seem to be true, and necessarily so, according to current cladistic uses of 'Mammalia' and 'Aves' (see Chapter 1, §I.3). All and only descendants of the right ancestor are members of Mammalia or Aves in any possible world. There could be other mammal-*like* or bird*like* organisms that evolve from a very different point on the tree of life. But these organisms would not belong to Mammalia or Aves. They would only be similar organisms produced by convergent evolution (see Figure 2.1).

In the same way, there is no possible world in which the descendants of mammals or birds in that world evolve into something else instead. Descendants might come to resemble insects, but they would not *be* insects. They would only be insectlike mammals or birds produced by convergent evolution.

So 'Mammalia = the clade that stems from the ancestral group $G$' and 'Aves = the clade that stems from the ancestral group $A$' seem true, and indeed necessarily true, at least given some prominent systematists' uses of 'Mammalia' and 'Aves'. But though necessarily true, 'Mammalia = the clade that stems from the ancestral group *G*' and 'Aves = the clade that stems from the ancestral group *A*' do not seem to me to have been *discovered* to be true. Earlier speakers did not use 'Mammalia' and 'Aves' as cladists use them today. Rather, I argue in the next chapter that these terms have undergone meaning refinement to *make* them refer to the relevant clades. This refinement has made 'Mammalia = the clade that stems from the ancestral group *G*' and 'Aves = the clade that stems from the ancestral group *A*' true.

'*Brontosaurus = Apatosaurus*', on the other hand, seems to have been discovered to be true, as I have already indicated. Riggs determined that '*Brontosaurus*' and '*Apatosaurus*' had all along been referring to the same dinosaur, though people had not realized that. He did not change the use of '*Brontosaurus*' or '*Apatosaurus*' to make the sentence '*Brontosaurus = Apatosaurus*' true; he just learned that the sentence was true and called people's attention to that fact.

So theoretical identities in general seem to me to be unlike '*Brontosaurus = Apatosaurus*' in that they are not discovered to be necessarily true. Even so, I defend in the following two sections the position that the theoretical identities discussed by philosophers resemble '*Brontosaurus = Apatosaurus*' in an important respect: They contain two rigid designators, just as Kripke says that they do. They need not be true, and even those that are true do not seem to have been discovered to be true. But because these identities contain two rigid designators, they are *necessarily true*, *if* true at all. That is what rigidity guarantees.

### I.7. *Descriptions and De Jure Rigidity*

The claim that theoretical identity statements are necessarily true in virtue of containing two rigid designators faces its own particular set of difficulties. The most prominent difficulty arises because the theoretical expressions that are supposed to describe essence do not seem to be proper names. Unlike 'Cicero' and '*Brontosaurus*', 'the element with atomic number 79' and 'the clade that stems from the ancestral group *G*' seem to be descriptions. And unfortunately, as Bolton (1996, p. 148) and Barnett (2000, pp. 108–9) point out, descriptions are unlike names in that they are not always rigid. Some descriptions happen to be rigid, such as 'the positive square root of 1600'. No number other than 40 is described in any possible world by 'the positive square root of 1600', because it is impossible that 1600 should have had a different positive square root. But although some descriptions are rigid in this way, others are not rigid. A nonrigid designator for 40 is 'the number that Kim would rather not think about as his birthday approaches'. Kim might have felt squeamish about a different number as his birthday approaches, and in some worlds he does.

Descriptions are sometimes rigid and sometimes not. Kripke calls descriptions like 'the positive square root of 1600' rigid de facto. He says that *names* are rigid for a different reason: Names are rigid de jure, or by *stipulation* (1980, p. 21n.). Assuming that '40' is a name of the number 40, '40' is rigid de jure.

Because some descriptions are not rigid, an argument is needed to show that apparent descriptions like 'the element with atomic number 79', '$H_2O$', and 'the clade that stems from the ancestral group $G$' are rigid, if their rigidity is to be established. If no good argument is available, then we should suspend judgment about whether expressions like the foregoing are rigid.

That is the problem, or at any rate the challenge. I think it can be met. I think that expressions like 'the element with atomic number 79', '$H_2O$', and 'the clade that stems from the ancestral group $G$' are rigid. Such expressions are rigid de jure, like names. Or at least they are rigid de jure if they are not rigid de facto.

Take 'the clade that stems from the ancestral group $G$'. I claim that 'the clade that stems from the ancestral group $G$' is a rigid designator for the clade whose organisms are, in any possible world, just those organisms belonging to $G$ or descended from $G$. But how can I rule out the possibility that the expression is nonrigid, so that in other worlds 'the clade that stems from the ancestral group $G$' applies to a different clade? Perhaps 'the clade that stems from the ancestral group $G$' could apply in *our* world to the clade that stems from $G$ (in our world), while it applies in some *other* possible worlds not to that same clade, which stems from $G$ in our world, but rather to another clade that in other worlds but not ours descends from $G$. Perhaps, further, the referent of 'the clade that stems from the ancestral group $G$' in *our* world has a different stem $S$ in some other worlds. In that event, 'the clade that stems from the ancestral group $G$' does not refer to the same entity in all possible worlds in which that thing exists, and otherwise to nothing else. It is not rigid.

The objection is, then, that possession of $G$ for a stem may not be a necessary and sufficient condition, in other possible worlds, for a clade's being identical to the clade that has $G$ for a stem in this world. Biologists' use of 'clade' suggests otherwise: When biologists discuss counterfactuals, they identify one clade across possible worlds just by its stem, possession of which is evidently supposed to be a necessary and sufficient condition in any possible world for being the relevant clade. If possession of a given stem is indeed a necessary and sufficient condition for being the clade associated with that stem, then the objection fails: 'The clade that stems from the ancestral group $G$' is rigid, because any clade in any possible world with $G$ for a stem is thereby the same clade as the clade that has $G$ for a stem in this possible world, and the clade that has $G$ for a stem in this possible world has $G$ for a stem in all other possible worlds in which it exists. Hence, 'the clade that stems from the ancestral group $G$' refers to the same clade in all possible worlds in which that clade exists, and never to anything else. Perhaps it is not obvious whether such rigidity is de jure or de facto, but that issue need not be resolved.

I have claimed that 'the clade that stems from the ancestral group $G$' is rigid if possession of a particular stem like $G$ is in any possible world a necessary and sufficient condition for being the clade that has this stem in the actual world. But what if I am wrong about the necessary and sufficient conditions for being a particular clade? Perhaps the use of 'clade' in the language as a whole is not as settled in the direction of the foregoing answers as I have supposed, for example. In that event, I will argue, 'the clade that stems from the ancestral group $G$' refers rigidly to the relevant group de jure, *by stipulation*, at least for biologists like those I have been discussing (again, see Chapter 1, §I.3). Biologists can just stipulate that they use the expression in various possible worlds for $G$ and $G$'s descendants. Nothing can stop biologists who wish to do so from coining a term to talk about that group. Scientists are free to pick an expression, say '$G$-clade', and just stipulate that it does refer to that group. If they do so, then '$G$-clade' *rigidly* designates, de jure, that group. So there is no worry that '$G$-clade' fails to designate the right group rigidly. But of course, if scientists could stipulate that '$G$-clade' rigidly designates the right group, they could stipulate that 'the clade that stems from the ancestral group $G$' does so, too. Or, scientists could just abandon 'the clade that stems from the ancestral group $G$' and use '$G$-clade' instead.

There is no worry, then, that the expression 'the clade that stems from the ancestral group $G$' does not rigidly designate the right kind. The same can be said for expressions like '$H_2O$' and 'the element with atomic number 79'. They are rigid de jure, at least if they are not rigid de facto, so there is no worry that they are nonrigid. Kripke hints that he would agree. He discusses a similar complication involving the expression 'C-fibers', which he takes to be rigid. He considers that 'C-fibers' might turn out *not* to be rigid, but he dismisses the problem quickly: "The point is unimportant; if 'C-fibers' is not a rigid designator, simply replace it by one which is, or suppose it used as a rigid designator in the present context" (1980, p. 149).

Given that 'the clade that stems from the ancestral group $G$' is rigid, and given that, in accordance with the arguments in earlier sections, 'Mammalia' is rigid, 'Mammalia = the clade that stems from the ancestral group $G$' is necessarily true if true at all. And 'Mammalia = the clade that stems from the ancestral group $G$' *does* seem to be true, at least in the idiolects of some prominent systematists (again, see Chapter 1, §I.3). Thus, the statement seems to be necessarily true.

### I.8. *Theoretical Identities and Changing Meanings*

Before concluding my discussion of rigidity, I address one more objection to the rigidity of expressions like 'Mammalia' and 'the clade that stems from the ancestral group $G$' and thus to the necessity of statements like 'Mammalia = the clade that stems from the ancestral group $G$'. The objection is from Ghiselin:

> [W]e scientists do not attach a name to a class, then discover the defining properties which are its essence, but rather redefine our terms as knowledge advances. Therefore the view of Kripke (1980) and his followers (see Schwartz, 1977) that natural kind terms are, like proper names, "rigid designators," should be dismissed as nugatory, and with it the accompanying essentialism. (Ghiselin 1987, p. 135)

I share Ghiselin's misgivings about the received view that scientists' conclusions about the essences of kinds they investigate are typically discoveries. I have already conceded that scientists have not discovered 'Mammalia = the clade that stems from the ancestral group $G$' to have been true given earlier speakers' use of 'Mammalia'. As Ghiselin suggests, there has been meaning change. Still, all of this is quite irrelevant to *rigidity*.

It is important to distinguish the position that 'the clade that stems from the ancestral group $G$' is rigid from the position that 'the clade that stems from the ancestral group $G$' describes the discovered essence of what speakers have all along been calling "Mammalia." For the present, let it be granted that the discovery thesis is false, and that our natural-kind terms are constantly being redefined, so that the precise kind that is referred to continues to change as the terms' meanings continually change. In that case, we should not expect a kind term to rigidly designate the *same* kind before the rise of modern science as it designates after the rise of science. But *at any given time* it could still rigidly designate a single kind: That is, at any given time in the history of a given language, a kind term might designate, in each possible world, the same kind, though the term may not at other times designate that kind because its meaning varies from time to time. So the mere evolution of meaning does not destroy the rigidity thesis. A term need not keep its meaning over time in order to be rigid at a time. Ghiselin's objection to rigidity does not tell against the present account of rigidity, according to which a kind term is rigid by virtue of designating the same kind in every possible world.

'Mammalia = the clade that stems from the ancestral group $G$' is true, given a certain use of 'Mammalia' adopted by particular scientists today. Indeed, I have argued that this statement is *necessarily* true. But that is not to say that the sentence was true in earlier times as earlier speakers used it. I doubt that it was, as I have said, though a proper refutation must wait until the following two chapters.

## I.9. *Nominal Essences or Real Essences?*

I have argued that 'Mammalia = the clade that stems from the ancestral group $G$' is true, given a certain use of 'Mammalia' adopted by particular scientists today. Further, I have argued that this statement is necessarily true and reveals the *essence* of the kind Mammalia. Species and higher taxa have scientifically interesting essences, just as the philosophical tradition inherited from Kripke and Putnam would have it.

But here some would call upon me to distinguish between kinds of essence.[12] Even John Locke, hardly a forebear to Kripke and Putnam, acknowledges that biological kinds have what he calls "nominal essences." Kripke and Putnam seem to affirm something more substantive: that biological kinds have "real essences," in Locke's terminology. Is my claim merely the Lockean one that biological kinds have nominal essences? No, it is not.

For Locke, a term like 'mammal' refers to whatever possesses the properties in the nominal essence associated by the speaker with the term. The nominal essence is an idea in the mind of the speaker. So for the Mammal kind, speakers might associate the properties *fur-bearing*, *lactating*, and so on with the kind. 'Mammal' refers in any possible world just to objects that have those properties. Clearly on my account 'mammal' does not refer just to whatever has properties like these.

On my account, whether or not 'mammal' refers to an organism is determined by whether or not that organism bears the right relationship to a certain ancestor that is specified by way of sample organisms. This is an account that Locke openly rejects. For Locke, the proper application of a kind term to a member of the kind cannot depend upon the member's relationships to objects or structures that are not familiar to the user of the term. Otherwise the term would be meaningless. Terms whose use is dependent on what is unfamiliar to a speaker "would be the Signs of he knows not what, which is in Truth to be the Signs of nothing."[13]

So on the account that I have given, the essence of a biological taxon is not merely "nominal," as Locke would have it. Nor is the essence of a biological taxon a Lockean "real essence," however. Locke's "real essences" are microstructural, not historical. The essences of taxa are historical. Even so, in order to specify essences for taxa, biologists make reference to scientifically interesting entities (ancestral groups) unfamiliar to them, just as they would if they were assigning Lockean real essences to a taxonomic term. For this reason, the essences of taxa have much of the character of Lockean real essences. Like real essences, they are theoretical entities unfamiliar to speakers: As Putnam would put it, they are "out of the head."

This externalistic position may initially appear to commit me to saying that taxonomic essences are discovered to belong to their respective taxa, but that is not so. I am committed to saying that it is an empirical and scientific matter whether any organism in doubt is indeed descended from $G$, and thereby a member of Mammalia. But I am not committed to saying that scientists have discovered the truth of statements revealing the essence of Mammalia, such as 'All and only descendants of $G$ are members of Mammalia' or 'Mammalia = the clade that stems from the ancestral group $G$'. I will soon argue (in Chapter 3) that scientists have *not* discovered the truth of such statements about essence.

## II.   ESSENTIALLY BELONGING TO A KIND

I have discussed necessarily true *identity* statements about kinds. But essentialist claims go beyond these. Many essentialists would say that, for example, the tiger species *Panthera tigris* is essentially a species of mammal, though of course *Panthera tigris* is not *identical* to Mammalia. And many would say that *individual* tigers that belong to *Panthera tigris* belong essentially to that species, as well as to Mammalia, Chordata, and so on.

Here in section (II), I address these common essentialist claims about kinds. In section (II.1) I discuss whether taxa of species rank and higher essentially belong to the higher taxa into which they are nested – for example, whether the species *Panthera tigris* essentially belongs to the broader kinds Mammalia, Chordata, and so on. In (II.2) I discuss whether the individual organisms and groups that belong to a *species* do so essentially – for example, whether Tabby the tiger essentially belongs to *Panthera tigris*. Interestingly, the answers to these two questions differ.

## II.1. *On Essentially Belonging to a Higher Taxon*

In this section, (II.1), I consider the essentialist claim that if a species or higher-ranked taxon belongs to a genus, family, order, and so on it does so essentially. I argue that this claim is a highly plausible one. This type of essentialism accords nicely with current, history-based systematics. Given plausible definitions for taxonomic names, anyway, essentialism is hard to avoid.

In section (I.6), I argued that 'Mammalia' designates, at least given certain professional systematists' use of that term, the clade whose members in any possible world are members of the ancestral group $G$ or descendants thereof. A species term such as '*Panthera tigris*' similarly designates a group originating in its ancestral population. The relevant population ancestral to the type specimen of the tigers may be given a name: "$P$." It could not happen that any members of *Panthera tigris* fail to be descended from $P$.

But is it necessary that the members of *Panthera tigris* also be members of Mammalia? Assuming the soundness of my argument (in §I.6) that there is no possible world in which an organism is *both* a descendant of Mammalia's stem $G$ *and* a nonmammal, the only way that it could fail to happen that the members of *Panthera tigris* are mammals is if, in some possible world, they are *not* descended from $G$. And because the members of *Panthera tigris* are descended from $G$ in the *actual* world, there is no possible world in which a member of the species is *not* descended from $G$, at least if a population is essentially a product of its ancestors.

And a population *is* essentially a product of its ancestors, for cladists. Cladists' talk of counterfactuals makes no sense if populations and other chunks of the phylogenetic tree are not essentially products of their ancestors. Cladists would insist, for example, that the tiger species could not become a species of marsupial, on the grounds that tigers originate from a different chunk of the tree of life. Because tigers originate from a different chunk of the tree of life, they could never be descended from marsupials, and therefore they could never *be* marsupials. The assumption is that chunks of the tree of life possess their ancestry essentially. The chunk from which tigers have descended could not have descended from marsupials: It is essentially a product of the nonmarsupials that actually generated it.

The doctrine that a population's origins are essential to it might be questioned by some philosophers, but the matter should not be controversial: It can be determined by stipulation if it is in question. Cladists are interested in naming groups whose boundaries are determined by evolutionary relationships. If cladists wanted to be explicit about their commitments, they could be. Cladists could define 'Mammalia' as follows: "Mammalia $=_{df.}$ the clade that stems from the ancestral group $G$" (cf. de Queiroz 1992, 1995). Then they could specify the use of '$G$' as follows: "'$G$' is a name for that particular chunk of the phylogenetic tree that happens to be the closest common ancestor of the echidna and the horse." A similar specification could assure that '$P$' is a name for a particular chunk of the phylogenetic tree. And a similar specification could assure that any chunk of the phylogenetic tree is individuated by its ancestry.

Given specifications like these, it follows that every species or higher taxon *essentially* belongs to every genus, family, order, and so on to which it belongs. *Panthera tigris* is essentially a species of mammal. It is essentially a species of mammal because $P$ is essentially a descendant of all of its ancestral populations, including $G$, which lies at the base of Mammalia. That makes $P$ essentially mammalian, given that 'Mammalia' is defined as specified above: "Mammalia $=_{\text{df.}}$ the clade that stems from the ancestral group $G$." Because in any possible world any member of *Panthera tigris* is in or descended from $P$, and because in any possible world anything in or descended from $P$ is descended from $G$, it follows that in any possible world if something is a member of *Panthera tigris*, it is descended from $G$. It is a mammal.

And so it is that *Panthera tigris* is essentially a species of mammal. Besides being essentially a species of mammal, *Panthera tigris* is also essentially a species of synapsid, of tetrapod, of vertebrate, of chordate, and so on. To be sure, tigers could evolve in such a way that they *resemble* nonmammals or even nonchordates, but they would still be mammals and chordates. As the cladistic biologist Mark Ridley says, if a chordate species were to develop six legs and a jointed exoskeleton, it "would not suddenly have been catapulted into the insects." Rather, it would still be a chordate species; "only certain taxonomists' ideas about how to recognize a chordate would have altered" (1989, p. 2).

## II.2.   *On Essentially Belonging to a Species*

I have argued that, given standard cladistic definitions for taxa, any species that belongs to a higher taxon (a genus, family, order, and so on) does so essentially. Similarly, any genus that belongs to a taxon of higher rank than itself (family, order, and so on) belongs to that higher taxon essentially. Species and higher taxa belong essentially to the higher taxa in which they are nested.

In the present section (II.2), I turn to the question of whether what belongs to a *species* belongs to it essentially. That would include, most saliently, individual organisms like Tabby the tiger. It would also include tiger subspecies (Siberian tigers, for example, form a subspecific branch of tigers) as well as various tiger populations, and so on. I will discuss individual tigers first, though what I say about individual tigers will have straightforward application to subspecies and populations. So I begin with this question: Do individuals like Tabby essentially belong to *Panthera tigris*? In providing an answer, I will leave behind the kinds of sweeping essentialist claims that I made in section (II.1). Given cladism, or indeed given any mainstream account of species, membership in a species is not always essential to members.

This anti-essentialism runs contrary to common essentialist convictions, of course. Many people hold the view that an individual can belong to a species only if being of that species is essential to it. Thus, for instance, Cocchiarella, who understands "natural kinds to include the various genera and species of plants and animals" (1976, p. 203), insists that "an individual *can* belong to a natural kind only if being of that kind is essential to it" (p. 205). Doepke (1992, p. 89) agrees that no member of a natural kind could become a member of a different kind. For instance, he says, "no human person could become a member of another species." The views expressed by these authors and many others do not accord with contemporary systematics.

II.2.a. ESSENTIALISM'S STATUS GIVEN CLADISM. Because I focus on cladism in the foregoing discussion about higher taxa, I begin the discussion of species with a look at the cladistic account. Orthodox cladists (e.g., Meier and Willmann 2000, p. 31; Ridley 1989) follow Hennig, the founder of the cladistic school of systematics, in maintaining that every splitting of a species marks the beginning of two or more new species and the end of the ancestral species. Ridley, who adjusts Hennig's definition slightly in order to allow for cases of extinction and living species, understands a species to be "*that set of organisms between two speciation events, or between one speciation event and one extinction event, or that are descended from a speciation event*" (Ridley 1989, p. 3). A *speciation event* is a branching off of a new species. Each species begins for cladists with a forking event and ends by being wiped out, or when it itself forks into separate species (see Figure 2.2).

A moment's glance reveals that the foregoing definition is incomplete. We need some account of speciation, of what counts as a splitting of species lineages. Then it will be understood what counts as the birth and death of a species. For example, cladists might say that a species splits when it becomes divided into two or more groups that are no longer capable of interbreeding with each other. Another plausible view might be that a species splits when it becomes divided into two or more groups with unique ecological niches. Whether one of these options or some other is chosen is not determined by cladistic principles.

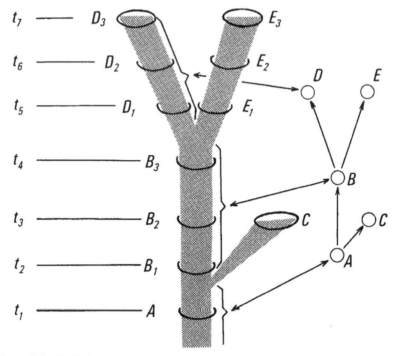

**Figure 2.2.** Cladistic species. From W. Hennig's *Phylogenetic Systematics*. Copyright 1966, 1979 by the Board of Trustees of the University of Illinois. Used with permission of the University of Illinois Press.

Ridley's intent is that a species name like '*Panthera tigris*' is to be defined something like as follows: "*Panthera tigris* =df. *the lineage descending from ancestral population* P *and terminating in speciation or extinction*," P being, as I indicate in an earlier section, an appropriate population in the lineage that gave rise to today's tigers.[14] Should a zoo manage to spawn a new species of tigers next year, *Panthera tigris* would end. Two new descendant species would be born.

Now, cladism might seem at first to be compatible with essentialism. After all, one's place in one's own species is, for cladism, entirely determined by one's place in a genealogical tree, which would seem to be essential, at least if Kripke (1980) is right that organisms are essentially products of their genetic parents.

But cladism is bound to disappoint the essentialist. According to cladism, a species becomes extinct whenever it sends forth a new side species. This is so even if the lineage undergoes no change after sending out the side branch, so that earlier members are indistinguishable from later ones. Thus, in Figure 2.2, species *A* becomes extinct upon the arrival of *C*, even though the lineage including *A* and *B* does not evolve at all. For this reason, as Kitcher points out in a different context and for different reasons (1989, pp. 200–2), whether or not a species like *A* becomes extinct at a particular time depends upon whether a side isolate threatening to branch off at that time is wiped out by a cataclysm before achieving species status.

These conditions for membership in a species make it the case that the members of some new species, such as *B* above, are not *essentially* members of that species. They would be members of the ancestral species instead, if only a side branch, *C* in this case, had never attained reproductive isolation, perhaps having been destroyed by some cataclysm first. Whether members of *B* belong to species *B* or species *A* depends on whether an accident occurs that ought to have nothing to do with whether those members exist.

One might try to hold on to the doctrine of essentialism by maintaining that had *C* been wiped out, the organisms that make up *B* would *not* have existed, even though other organisms that are intrinsically just like the members of *B* and that occupy the same spatio-temporal location as the members of *B* would have existed. But this is hardly palatable. It commits the essentialist to saying that if some random speciation event occurs in the line of *Homo sapiens* in the next ten minutes in Wyoming, you will cease to exist! There will be a breathing, thinking organism in your place, one that is spatio-temporally continuous with you, one that is composed of the very same particles of which you are made the moment before you go out of existence, one that looks like you, talks like you, acts like you, responds to your name, and so on, but this will be an entirely different individual. *You* will have gone out of existence. As tough as this position is to swallow, the essentialist has no other choice. If she rejects it and says that you will still be around in ten minutes, then she must concede that you could change species, by cladistic standards. That is because even though you will still be around, *Homo sapiens* will not. It will have ended, and given rise to two new species, one of which contains you as a member.

Occasionally people to whom I have presented this case against essentialism have suggested to me on *behalf* of essentialists that essentialists might try to say that if your species goes out of existence, so do you. But no essentialist has ever claimed this as her *own* response. The reason seems clear. Essentialists by and large *agree* that no speciation event in Wyoming in the next ten minutes would terminate you. They agree that even if in ten minutes some feat of genetic engineering in Wyoming results in some offspring of humans being members of a different species incapable of interbreeding with *Homo sapiens*, you will still exist. The reason essentialists have been essentialists is not that their intuitions do not accord with this conclusion. Rather, the reason they have been essentialists is that they have not realized that their position commits them to saying otherwise. Essentialists have simply been uninformed about systematics.

II.2.b. ESSENTIALISM'S STATUS WITHOUT CLADISM. Placement in a particular species is apparently not essential to an individual, at least on a cladistic account of species.[15] But what if we abandon cladism? There is an interbreeding conception of species that is *not* cladistic. This approach is most famously exemplified in Mayr's biological species concept (BSC). According to the BSC, species are "groups of interbreeding natural populations that are reproductively isolated from other such groups" (Mayr 1969, p. 26; 1970, p. 12; 1976, p. 518).

Cladism really represents a modification of Mayr's concept or some other. The difference between the BSC and the cladistic account that makes use of it is in where species begin and end. The BSC dictates endpoints that are determined not by branching, but rather by potential interbreeding. Obviously the ancestors and descendants of a particular lineage are precluded by temporal barriers from breeding with its present members, but what matters for determining conspecificity is that they *would* be capable of interbreeding *were* they contemporaries. For the BSC, species *A* in Figure 2.2 would not become extinct after the budding of side branch *C*; rather, there would be no species *B*. What Hennig counts as the members of a new species *B*, Mayr would count as members of *A*: After all, because no evolution has occurred from time $t_1$ to time $t_4$, the members of the two groups would be capable of interbreeding.

Similarly, if there were a very long lineage that had evolved a great deal over time but that had never happened to branch, so that earlier members of the lineage were incapable of breeding with later members, Mayr would break the lineage into two or more species. Cladists would not: All members of the lineage are conspecific so long as no speciation event severs them. The endpoint of a species must be set by splitting for cladists.

Because splitting is what was shown previously to cause trouble for essentialists, one might hope that the BSC itself, unmodified by cladism, would support essentialism. The same might be hoped for other species concepts, when they are left independent of cladism. The BSC is closely related to Paterson's (1985) recognition species concept, for example, on which a species is a group of organisms with a shared mate recognition system. According to the ecological approach, endorsed by Van Valen (1976) and Andersson (1990), a species is a lineage with a unique adaptive zone, or ecological niche.

But cladism is not needed to make trouble for the position that the members of species are essentially members. Both the interbreeding approach and the ecological approach seem to distinguish species on the basis of features that are only contingently possessed by organisms. Hence, species membership or nonmembership would not seem to be essential if either account is right. Consider a large population of organisms, from which a small population splinters off and takes up a new ecological niche, adapting to a new way of life. Organisms of the two branches cease to be recognizable to each other as mates and thus become reproductively isolated. Both of the foregoing approaches would consider the two branches to be distinct species. Organisms of the smaller branch do not belong to species "*A*," whose members constitute the larger branch. Yet this could be a plainly contingent matter. Had the members of the little branch not taken on a new niche, or had there not been reproductive isolation (or failure of mate recognition) between them and the members of *A*, they would belong to species *A*, given the species concepts in question. Members of the little branch are not essentially such that they belong to a new niche, or that they are reproductively isolated, as I will explain.

Surely it is possible that members of the side branch should have remained in the original niche. Taking on a new niche might have involved such alterations as a change in diet or in predators. But clearly the members of an isolate like the foregoing are not essentially such that their lineage has the diet or predators it has.

Likewise, it is certainly possible that the side branch should have failed to become reproductively isolated from members of the main branch. This might seem less clear than that the members could have taken on a new ecological niche. It might seem less clear because if reproductive isolation obtains, some isolating mechanism, more than just geographical or temporal distance, prevents interbreeding. The most likely isolating mechanism is genetic in nature: Because of some mutation in the isolate, there can be no interbreeding with the parental species. If that is the reason for the reproductive isolation, then many essentialists might be expected to insist that members of the side branch are essentially members of a different species, because it would take a genetic change, albeit perhaps a very minor one, for them to be members of the original species. Although Kripke (1980, p. 115n.) hints that some minor genetic change could occur to an individual without that individual's losing its identity, the issue raises unwelcome complications. So rather than defend the possibility that an individual in the isolate could have had a slightly different genetic makeup, I will point out that genetics may have nothing to do with reproductive isolation.

A lineage could become reproductively isolated even though its members do not change genetically at all. This could happen in a number of ways, including by way of learned habits that prevent crossing. As Ghiselin says, "speciation can occur without genetical change, as when isolating mechanisms are the result of learned behavior" (1987, p. 137). This learned behavior might make for speciation between groups that are slightly different from one another but that might have continued to interbreed anyway, and hence to remain conspecific, if one or both groups had not happened to learn to avoid interbreeding.

Or consider the reproductive isolation that obtains between two lineages of flowering plants on account of insect pollinators' preventing crossing between the lineages. It is hardly essential to any individual plant that its lineage be or not be crossed with another lineage because of local insects' behavior. But this could make the difference between whether a plant belongs to one species or another. Two strains of a plant may differ a little in various ways but still be crossed by an insect pollinator. One strain of plants may, for example, flourish in acidic soil, while the other may be intolerant of such soil. The differences could be genetic, even though the strains are conspecific because they are crossed by insects.

Suppose the variety of insect that crosses the two strains is wiped out and replaced by other insects that do not cross the strains. The new insects might be more specialized, so that a different type of insect prefers to tend exclusively to each strain of plants. If this were to happen, the strains would become reproductively isolated. They would cease to belong to the same species. The individual plants of one strain, a small isolate, we may suppose, would become members of a new species. But it would be a contingent matter that they ever did. The original insect pollinator might never have been supplanted by others, and thus the members of the isolate might not have become members of a new species.

For the foregoing reasons, it certainly seems as if the BSC allows for contingent membership. So does the mate-recognition concept. It seems possible that the organisms of two branches that cease to recognize each other as mates should have continued to recognize each other as mates. Being such that one recognizes these as opposed to those as mates would not seem to be an essential property, as the foregoing example of plants suggests. There are other examples: Many frogs use distinct calls to recognize mates, but it is unlikely indeed that any frog should essentially have the call it has, especially because the acoustic properties of its call depend on variables such as the temperature of the surrounding environment.

Organisms in a new isolate that take on a new niche, or become reproductively isolated, or cease to recognize members of the stock branch as mates might not have done any of these things. The organisms of the isolate happen to constitute a unique species, $B$, but membership in $B$ is a contingent matter for the foregoing accounts of species; things could have gone in such a way that these organisms were not members of $B$, being members of the stock species $A$ instead. Hence, on the interbreeding and ecological accounts of species, organisms can be contingently included in, or excluded from, a species.

There are many currently proposed accounts of species other than the interbreeding and ecological accounts, as I indicate in the next chapter. It would seem that the other standard accounts generally have their own unique features that put them at odds with essentialism, just as the interbreeding and ecological accounts do. But I will not run through the particularities of a great many accounts here. Rather, now that I have shown that the *particularities* of some of the best-known accounts of species set these accounts against essentialism, I will argue that certain features shared by *all* standard accounts set them all against essentialism. I will argue that every account that so much as allows for the possibility of evolution, as all mainstream professional accounts on the market do, also allows for the possibility that organisms do not essentially belong to their respective species.

The more general argument against essentialism is best shown by way of an example adapted from Kitcher (1989, pp. 202ff.), though Kitcher uses the example to make a point very different from the one I wish to make, one that has nothing to do with essentialism. Kitcher imagines (see Figure 2.3) an evolving population that divides at $t$ into equal halves. Suppose that the members of the stem population constitute a species, by whatever criterion might be employed: interbreeding, ecological niche, or even similarity of some variety or other. By $t'$ the branches have diverged sufficiently to constitute separate species, whatever the criterion used to determine that: reproductive isolation, or something else. The divergence stops at $t'$. At no time is the distance between the ancestral lineage and either branch sufficient for speciation; speciation occurs only because of the distance attained between the two branches.

One is not sure what to say in such a case. $S_1$ and $S_2$ are stages of different species. It might be said that $S_0$ and $S_1$ are conspecific, as are $S_0$ and $S_2$, though this is odd, given that $S_1$ and $S_2$ are not conspecific. Perhaps we can live with the oddity of $S_0$'s belonging to two species. But if so, then it cannot be said that membership in any given species is an essential property. Members of $S_0$

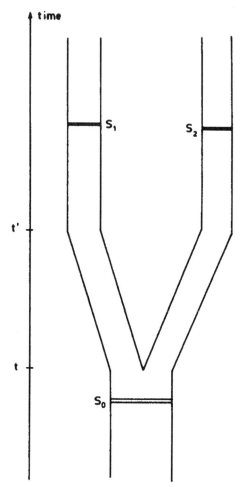

**Figure 2.3.** A speciation event. From P. Kitcher's "Some Puzzles About Species," in M. Ruse (ed.), *What the Philosophy of Biology Is: Essays Dedicated to David Hull.* Copyright 1989 by Kluwer Academic Publishers. Used with kind permission from Kluwer Academic Publishers.

might not have belonged to both species of which they are members. They might have belonged to only one species. They *would* have, had one branch been wiped out before speciation could occur.

Another option is to declare the ancestral species to be extinct at *t*. This, too, ill accords with essentialism. Members of each branch would not necessarily belong to a species other than the ancestral species. A cataclysm could have made things otherwise: For had either branch been prevented from evolving until $t'$, there would not have been sufficient distance between the branches for speciation to have occurred. Kitcher's third solution involving pluralism provides no lifeline for essentialism: It only recognizes both of the foregoing approaches, counseling biologists not to insist on one solution or the other. It looks as if the matter cannot be resolved in a way that allows membership in the species to be essential.

Essentialism is, then, in some trouble. No current account of species can tolerate it. Granted, I have considered only current accounts of what species are and current views about how species behave. Though current biology is ill disposed toward essentialism, there is always the possibility that the future of biology will vindicate essentialism. But, of course, if essentialism is forced to bank upon this hope, it is undergirded by little support indeed. Essentialism of the relevant variety is embarrassingly at odds with current, mainstream systematics.

II.2.C. ESSENCES OF SPECIES VIS-À-VIS ESSENTIALLY BELONGING TO SPECIES. I have argued that individual tigers can fail to belong essentially to *Panthera tigris*, and by a straightforward extension of that reasoning, we should conclude that populations and subspecies in *Panthera tigris* can also fail to belong essentially to *Panthera tigris*. But it may be worth clarifying that this is not to say that the *species Panthera tigris itself* has no essence. It does have an essence, just as higher taxa like Mammalia do.

Whether it is correct to say that the essence of *Panthera tigris* is this or that depends on how we use '*Panthera tigris*'. We might use '*Panthera tigris*' as cladists do, or we might use it as other systematists do. According to each use of the term, there is an essence for the kind named. I will look at just the cladistic use, to illustrate the point.

Recall the cladistic definition for '*Panthera tigris*': "*Panthera tigris* =df. the lineage descending from population* P *and terminating in speciation or extinction.*" According to that understanding of *Panthera tigris*, the species has an essence. The essence is to be the lineage that descends from *P* and that terminates in speciation or extinction.

Given the foregoing definition, every possible world is such that all and only the organisms belonging to *Panthera tigris* belong to the relevant lineage descending from *P*. There is no world in which anything that belongs to *Panthera tigris* fails to belong to that lineage, and no world in which anything that fails to belong to *Panthera tigris* belongs to that lineage. True, some members of *Panthera tigris* are not members of *Panthera tigris* in every possible world in which *those members* exist, so they do not belong to the relevant lineage in all possible worlds in which they exist. But those members

61

do belong to the relevant lineage in just those worlds in which they happen to be members of *Panthera tigris*. They have to, because although it is not essential to everything that belongs to *Panthera tigris* to belong to the relevant lineage, it is essential to *Panthera tigris* itself to *be* that lineage. *Panthera tigris* is that lineage in all possible worlds: To be that lineage is its essence.

## III. CONCLUSION

The fate of various essentialist theses must be distinguished. Some theses seem to be wrong, in light of current biology, but others seem to be right. Although the individuals, subspecies, and so on that belong to a particular species might be such that they do so only contingently, species belong essentially to the higher taxa in which they are nested, at least given standard cladistic definitions for taxonomic names. Something similar can be said about taxa of higher rank than species and the still higher taxa in which they are nested. And species themselves have essences, even if possession of the traits that are essential to a species is not essential to all members of the species. These were the points made in section (II). In section (I), I argued that kinds have essences that are exposed by necessarily true theoretical identity statements. I also argued that some necessarily true identity statements about kinds are not theoretical identity statements, which expose essence; some necessarily true identity statements about kinds contain only ordinary names for kinds, and no essence-exposing designators.

A healthy portion of essentialism about kinds withstands scrutiny. The species and higher taxa recognized by contemporary biologists seem to have scientifically interesting essences. Because species and higher taxa seem to have scientifically interesting essences, it may seem reasonable to follow Kripke and Putnam, and a massive tradition following Kripke and Putnam, to the position that scientists' conclusions about kinds' essences are *discovered* to be true by empirical investigation.

But it would be a mistake to follow Kripke and Putnam, along with so many others, to this familiar position about scientists' conclusions. The received wisdom with respect to this position is mistaken. This is the subject to which I turn next.

# 3

# Biological Kind Term Reference
# and the Discovery of Essence

According to the received wisdom, scientists inquire into the nature of kinds that were named before much science was known. When all goes well, scientists discover the essences of the kinds in question. Scientists' conclusions about essences are discoveries, not stipulations. When scientists have discovered enough about the essence of a kind, they can correct past speakers who supposed that certain objects did or did not have what it takes to belong to the kind.

The foregoing picture is not initially implausible. Consider scientists' study of the rodents. Any schoolchild acquainted with the animals traditionally counted rodents could confirm the striking morphological and behavioral similarities between these animals. Nevertheless, many scientists now claim that some so-called "rodents" are impostors. Teams of scientists have presented molecular data over the past several years in support of the conclusion that guinea pigs, long counted rodents, have an evolutionary history that excludes them from the rodent camp.[1]

Supposing that these scientists' empirical information about evolutionary relationships is solid, scientists would appear to have discovered that guinea pigs are not rodents. That guinea pigs are not rodents would appear to be an empirically discovered necessary truth. In the same way, our claim that whales are not fish appears to have been discovered to be true. Scientists appear to have corrected past speakers who said otherwise. Empirical research has resulted in scientists' "discovering that 'whales are mammals, not fish' is a necessary truth" (Kripke 1980, p. 138). The frequency of such conclusions about natural kinds testifies to scientists' success as they endeavor, in the words of Penelope Maddy, "to discover the underlying traits that make these things what they are" (1990, p. 39).

Kripke has made the foregoing general picture of essence discovery familiar. He writes, "In general, science attempts, by investigating basic structural traits, to find the nature, and thus the essence (in the philosophical sense) of the kind" (1980, p. 138). Putnam has also done much to popularize the view (1975e). Kripke and Putnam insist that scientists do not *replace* an older usage of a term with a newer, more sophisticated one. Rather, they shed light on the way speakers have *all along* been using the term: "[S]cientific discoveries of species essence do not constitute a 'change of meaning'; the possibility of such discoveries was part of the original enterprise" (Kripke 1980, p. 138; see also Putnam 1975e, pp. 224–5).

Now, the account popularized by Kripke and Putnam is somewhat biologically uninformed, and this makes it vulnerable to premature rejection. But it should not be rejected for the wrong reasons. Kripke and Putnam wrongly suppose that chromosome structure (Putnam 1975c, pp. 141–2; 1975e, p. 240) or at any rate "internal structure" (Kripke 1980, pp. 120–1) is what binds the members of a biological kind like a species into a common kind. In general, as I have already emphasized, biologists do not delimit species and other taxa on the basis of intrinsic properties like these. Biologists generally place organisms into taxa on the basis of shared ancestry, as the foregoing example of the guinea pig suggests.

So Kripke and Putnam have made a mistake about what criteria biologists use in delimiting kinds. This error about how to characterize the essences of taxa can nevertheless be easily remedied, as I have suggested in earlier chapters. Given that biological kinds are delimited historically, the essences of kinds simply become historical. As Kornblith (1993, p. 111, note 10) observes, "there is no a priori reason why historical properties, such as evolutionary history, may not feature prominently in an account of the essential properties of a kind."[2] Kripke and Putnam would have committed to such evolutionary essences had they known more about biological systematics, presumably: As it is, Kripke (1980) recognizes historical essences for individual organisms.

Given historical essences, the Kripke-Putnam claim about essence discovery becomes this: Scientists aim to discover the hidden historical bonds that make the members of biological kinds members of those kinds. Historically delimited essences are as good as microstructural ones, so far as the general picture of essence discovery is concerned. Biologically informed literature suggests the foregoing picture of essence discovery, as Griffiths points out, by stressing that although past systematists have only dimly seen nature's real divisions, "The exact boundaries and the real basis of these divisions is revealed by evolutionary theory and modern systematics" (1994, p. 210).

The idea that scientists discover the historical essences of kinds named by people ignorant of evolution has been plausible, at least on its face, since Darwin. Darwin himself encourages the idea, writing that "community of descent is the hidden bond which naturalists have been unconsciously seeking" (1859, p. 420). Superficial properties do not determine kind membership; history does. "No one regards the external similarity of a . . . whale to a fish, as of any importance" (Darwin 1859, p. 414). Rather, if essences are historical, then the superficial similarities between whales and fish are only misleading, just as Kripke and Putnam hold: "[S]uch resemblances will not reveal – will rather tend to conceal their blood-relationship to their proper lines of descent" (Darwin 1859, p. 427).

The Kripke-Putnam view that scientists' conclusions about the essences of biological kinds are discoveries is not threatened by the historical nature of species and other kinds. Whether the view in question is *true* is another matter. I will argue that it is not true. Scientists' conclusions about the essences of our species and other groups are not discovered to be true. That is *not* because the taxa are historical; rather, I will make in the next chapter a similar point about the claim that scientists' conclusions concerning the essences of *non*historical chemical kinds are discoveries. Scientists' conclusions about the essences of kinds named by earlier speakers are not, in general, simply discovered to be true, whether those essences are historical or not.

The structure of the present chapter is as follows. In section (I) I address the claim that scientists correct past speakers' erroneous uses of kind terms. Scientists are supposed to make such corrections when lay speakers have accidentally placed organisms into groups that are unnatural from a biological perspective. I argue, on the contrary, that speakers' entrenched judgments about what belongs in the extension of a kind term are not, in general, subject to empirical refutation. Rather, revisions in speakers' conception of what belongs in the extension of a common word marks the replacement of an older, less precise use of the word with a newer, more sophisticated one.

In sections (II) and (III) I ignore correction and what happens when speakers have accidentally placed organisms into groups that are unnatural from a biological perspective. I look instead at cases in which speakers have correctly identified the members of a natural kind. I investigate the claim that scientists' conclusions about the essences of the relevant kinds in such situations are discoveries. Again, I find fault with the dominant claim. Scientists do come to new conclusions about the essences of traditionally recognized kinds frequently enough, but their conclusions reflect a substantial measure of fiat, following a strain to shape the use of old terms in the light of new findings.

In section (IV) I take a special look at vernacular terms. The lessons of previous sections have more obvious application to scientific terms than to vernacular terms. But I argue that they apply to both.

## I. DO SCIENTISTS CORRECT THE MISTAKES OF PAST SPEAKERS?

There is, to be sure, *something* importantly true about the Darwinian claim that scientists discover the hidden bonds uniting a group. For any three organisms $x$, $y$, and $z$, scientists can often discover whether $x$ and $y$ belong to a clade or species that excludes $z$. (Again, a clade is, roughly, a species and all of its descendants.) It may be that guinea pigs do not belong with mice and rats in any clade that excludes horses and other animals that no one counts as a rodent. In that case, the traditional rodents have turned out to be unrelated, at least by the historical standards of cladism, the dominant school of systematics.[3] The narrowest clade containing guinea pigs as well as mice and rats is much broader than the traditional rodents: It includes horses, seals, and primates, because guinea pigs are more closely related to them than to mice and rats.

Scientists discover historical relationships like the foregoing. What they do *not* discover is that earlier speakers, who knew less about the essences of guinea pigs and rodents, were wrong about whether guinea pigs are "rodents." Scientists may *appear* to have discovered that "the guinea pig is not a rodent." But was this conclusion discovered to be true? No, it was not. The conclusion that "the guinea pig is not a rodent" was not forced upon scientists. Scientists might have continued to insist that "guinea pigs are rodents," in light of the foregoing empirical information and been no more wrong or right about the essence of "guinea pigs" or of "rodents."[4]

Guinea pigs would still merit the label 'rodent' according to a couple of reasonable responses to the discovery that they are not as closely related to mice and rats as people had thought. After things called by a label like 'rodent' are found not to constitute an exclusive clade, scientists generally have two options for adjusting the use of the label in order to assure that the term is assigned a scientifically recognized clade. They can pare the unacceptable taxon down or they can extend it. They can either say that some things formerly believed to belong to the extension do not belong to it, or they can say that some things formerly believed *not* to belong to the extension belong to it after all. Scientists who have urged that the traditionally recognized rodents do not form a historical group have suggested that guinea pigs be ousted from the rodent camp. They have suggested paring down. But extending would have been another option. Scientists could have urged that in fact the "rodents" are far more inclusive than was previously realized and include horses, seals, and primates. On this resolution, we are all "rodents"!

This may seem incredible. But it is not uncommon for the extension of a term to be so drastically expanded beyond the paradigms originally used to anchor it, after empirical investigation. For paradigmatic dinosaurs we turn to the likes of *Tyrannosaurus*, *Stegosaurus*, and the long-necked, long-tailed giant *Apatosaurus* (or *Brontosaurus*). Nevertheless, cladistic systematists are now counting sparrows and ducks as less frightening models of the dinosaurs. Because birds evolved from the dinosaurs, they *are* dinosaurs. Dinosaurs are not extinct after all![5] Just so, scientists might have concluded on the basis of the guinea pig study that early rodents split and evolved along different lines; but all of the products are evolved rodents, just as birds are evolved dinosaurs.

Other examples of extension abound. Some years ago Simpson (1961, pp. 121–3) supposed the mammals to have arisen from more than one reptilian ancestor but urged that, rather than declare the mammals nonhistorical, the systematist might group formerly supposed reptiles into Mammalia. In this way the first common ancestor of the would-be reptile lineages that separately gave rise to the mammals would be the first mammal.

Guinea pigs could have been counted "rodents," because scientists could have said that there are more "rodents" than they had thought, rather than fewer. There were two options. Neither option seems to have been forced. The headline-making conclusion that "the guinea pig is not a rodent" was not discovered to be true, and scientists would not have been mistaken to have concluded otherwise. More precisely, neither the conclusion that guinea pigs are "rodents" nor that they are not "rodents" is quite right or quite wrong on the earlier use of 'rodent'. The earlier use is vague about the matter. To make one conclusion standard or correct, the meaning of 'rodent' had to be altered. Scientists would have been entitled to alter language either way, so neither possible conclusion seems to represent a discovery about what have all along been called "rodents." It is not as if one conclusion gets the facts right and the other gets them wrong.

Whether scientists had said that there are fewer "rodents" than they had thought or more of them, 'rodent' would have remained tied to a scientifically respectable group. There is yet another natural conclusion that would have allowed scientists to continue calling guinea pigs "rodents," in this case *without* saying that the extension of 'rodent' is different from what earlier speakers took it to be. Scientists could have concluded that the "rodents" are just the animals traditionally called "rodents," including guinea pigs but not primates, by saying that studies on the evolutionary history of these animals have shown that the "rodents" are *not a historical group* worthy of taxonomic recognition.

Contemporary systematists are used to surrendering the taxonomic status of various terms like this in the light of phylogenetic evidence about extensions. Thus, systematists have acknowledged that 'algae' has no taxonomic correlate, because algae are supposed to share their closest common ancestor with, for example, horsetails and clubmosses.[6] 'Reptile' is a similar term. Modern cladistic systematists have abandoned 'reptile', concluding that it fails to designate a "monophyletic" group, which is to say (in the language of contemporary cladistic systematics) that the reptiles do not constitute a clade. The reptiles have given rise to animals not traditionally counted reptiles, so that any clade including the reptiles includes other organisms too.[7]

There are questionable terms whose fate is to be determined by further research. An example of one of these is 'zebra'. There is some reason to think that the closest common ancestor of the zebras might be an ancestor of our domestic horses; if so, then, as Gould notes, striped horses may merit the popular term 'zebra' by virtue of their striking similarities, but they are not a historical group worthy of cladistic recognition (1983, p. 358).

Just as 'zebra' and other terms may be released from taxonomic duty, 'rodent' could have been released. One interpretation of the discovered non-monophyly of the things that we have been calling "rodents" is that the "rodents" are just non-monophyletic. A "rodent" is the sort of furry, gnawing creature sporting those traits that used to mislead systematists into placing them in a common mammalian order, period.[8] So here is another way to resist the conclusion that "guinea pigs are not rodents." A term like 'rodent' might meet a variety of fates after phylogenetic disruption. Which fate attends it is for the working taxonomist to choose.

In the same way that 'rodent' could have been declared a nonscientific term in the face of disruption, so too might 'fish' have been declared a nonscientific term when speakers began to realize that they had been calling a few mammals "fish." In this way, speakers could have continued to say, in light of the same empirical evidence, "Whales are fish." Had that happened, experts would have said things like this by way of educating the laity: "The kind *fish* is not natural, because both whales and salmon are fish. From a biological perspective, whales should be grouped with other mammals to the exclusion of gilled fish."

Today the sentence 'Whales are fish' seems to express a falsehood, at least strictly speaking. Scientists and ordinary speakers alike refuse to call whales "fish." Occasionally children call whales "fish," but they are promptly corrected. They are told that whales are mammals, not fish. Because scientists and lay speakers do not call whales "fish," it seems that whales do not belong to the extension of 'fish'. After all, speakers are masters of their own words, and unless they lack crucial empirical information about sea creatures that is presently unknown, it is hard to see how speakers could be wrong in saying that 'fish' does not designate whales.[9]

68

The issue of error is a different matter. Were earlier speakers' statements that "Whales are fish" wrong? Have we just learned better? On the contrary, it seems much more likely that earlier speakers meant something different by 'fish'. Indeed, just what 'fish' should refer to was a question earlier authorities explicitly addressed. In the eighteenth century Oliver Goldsmith pondered that some groups recognized by earlier naturalists and by most ordinary speakers of his time had been shown to have little significance for naturalists. This raised questions about the continued use of terms for those groups. Was the whale a "nonfish" that had been misclassified by the general public and by earlier naturalists, or was the whale a bona fide member of the heterogeneous extension of 'fish'? After reflection, Goldsmith concluded that the whale is a lofty type of "fish." Goldsmith was not overlooking whales' distinctive character. Whales, he acknowledged, are "many degrees raised above other fishes in their nature" (1791, vol. 6, p. 167). But given the precedent of calling whales "fish," and a reason for that precedent, provided by the obvious similarities between whales and other so-called fish, Goldsmith determined that it was best to continue to call whales "fish": "[I]t is best to let them rest in the station where the generality of mankind have assigned them; and as they have been willing to give them all from their abode the name of fishes, it is wisest in us to conform" (1791, vol. 7, p. 1).

Naturalists' use of 'fish' did end up narrowing, in spite of Goldsmith's case for conservatism, and common use changed to mirror it. Still, Goldsmith's advice to the naturalists of his time seems sensible enough. Certainly naturalists could have agreed that whales are "fish," so that some mammals are "fish," given the precedent set by earlier speakers who had called whales "fish."

This idea rubs many people the wrong way. That our terms are to be mapped to scientifically recognized natural groups may seem inevitable, in view of the ubiquity of revision. But various terms testify otherwise. Sometimes, as I have shown, terms continue to be used for groups that are found, despite the intentions of earlier speakers, to be unnatural by contemporary, cladistic standards. I have already illustrated this point. Yet another illustration is found in the term 'lizard'. To make 'lizard' refer to a monophyletic group, snakes would have to be counted as lizards. Scientists have decided, rather than to depart from or revise earlier use so drastically, to abandon the use of 'lizard' as a natural-kind term. Cladistic systematists still use the term 'lizard', but with the clarification that it stands for an artificial group. Speakers do not always adjust the use of terms when they are found to designate unnatural groups, like the group consisting of whales plus gilled fish. And it is no error not to so adjust the use of such terms.[10]

## II.  ESSENCES AND COMPETING SCHOOLS

When speakers have accidentally applied a term to an unnatural group, scientists may adjust the use of the term, after investigating the matter. Scientists may, for example, conclude that some organisms traditionally taken to belong to the extension of a term do not really belong to the extension of the term. But such scientific conclusions are not discoveries. The alleged correction of past speakers' mistakes in naming seems dubious. That is the lesson of the previous section. I now put aside the issue of correction, along with situations in which a term has accidentally been applied to an unnatural group. I turn to cases in which scientists find that speakers have managed to apply a term just to related organisms that constitute a natural group. It might appear that in *these* cases scientists discover the historical essence of a kind.

Consider the birds, or the mammals. The organisms that speakers have called "birds" constitute a natural, historical group, and the organisms that speakers have called "mammals" constitute a natural, historical group. Something similar holds for various species of birds or mammals: *Icterus galbula* and *Panthera tigris*, for example. Have scientists discovered the essences of such groups?

Certainly scientists have been forthright about *assigning* essences to groups like these.[11] The question is whether scientists' essence assignments have been *discovered* to be true. I will argue that they have not. The most salient problem with the position that these essences have been discovered is that different systematists assign different essences to taxa, and it does not appear that any one systematic camp is right and its competitors wrong in this assignment. I elaborate on this problem of competition in the present section, (II). This is only one of the problems afflicting the position that the historical essences at issue have been discovered. That position would fail even without the problem of competition between competing schools, as I show in the following section, (III). In both sections, (II) and (III), I discuss both higher taxa and species: In the present section, I begin by discussing species.

## II.1.  *Species*

According to Kripke, Putnam, and a great tradition following them, conclusions about biological essences do not change the meanings of the relevant terms. Species terms' use is supposed to stay the same over time. I will argue that they do not. Again, there is meaning refinement. This is a change in terms' use.

Centuries ago, naturalists or ordinary speakers might have baptized 'tiger' or '*Panthera tigris*' by pointing to a shipment of tigers from Bengal and saying that the term is to refer to "the species to which *these* organisms belong." Because the tigers ostended in the baptismal ceremony would have belonged to a single species, such a baptism would have been successful. Even so, the essence of the group would not have been clear.

In time, the lack of clarity over the essence of the ostended kind would have become apparent: After some exploration, speakers would have become acquainted with closely related but distinct organisms like Siberian tigers, which are bigger, have thicker fur, and so on, and Sumatran tigers, which are smaller. To give an idea of differences between these distinct strains, an adult male Siberian tiger typically weighs between 500 and 700 pounds. An adult male Sumatran tiger weighs around half that, averaging between 250 and 300 pounds. The Sumatran tiger is much smaller. It also has other distinctive marks. For example, it has more closely spaced stripes, and longer whiskers. Its unique physical characteristics are adaptations to help it to survive in the dense jungle.

A question arises: Is '*Panthera tigris*' a name for all tiger strains, or only some of them? I submit that the foregoing speakers would not have done enough to provide a determinate answer to this question. Their baptism would have left the matter open.

You may be skeptical. After all, speakers did baptize '*Panthera tigris*' as a term for the *species* instantiated by the Bengal tigers. So if Sumatran tigers are of that species, then they possess the essence of *Panthera tigris*. Otherwise, they do not.

The problem with this response is that the boundaries of the *species* are not clear. Whether there are two species or just one present depends on what a species is. Scientists are supposed to settle the issue of what makes a species a species, and anyone familiar with the literature on species will be well aware that scientists are eager to pronounce on this matter. The question is whether scientists' pronouncements have been or will be *discovered* to be true. I will argue that the answer is likely to be no. The recent literature offers dozens of different professional conceptions of what a species is, and there does not seem to be any fact of the matter about which ought to be kept. It seems likely enough that one or another "species concept," as biologists call them, will prevail over others in biological discourse, but not that it will be discovered to be the true concept. This is a problem for the view that scientists' conclusions about the essences of particular species are discoveries.

Which of the different proposals for defining 'species' wins will determine whether the foregoing two strains of cats are conspecific. For example, according to Mayr's so-called biological species concept (BSC), species are "groups of interbreeding natural populations that are reproductively isolated from other such groups" (Mayr 1969, p. 26; 1970, p. 12; 1976, p. 518).[12] According to Cracraft's increasingly popular phylogenetic species concept (PSC), "*A species is the smallest diagnosable cluster of individual organisms within which there is a parental pattern of ancestry and descent*" (1983, p. 170). These two conceptions *both* divide the world into *natural groups*, but often these groups are not the *same*: The groups recognized by the BSC tend to be more inclusive than the narrower groups recognized by the PSC. Because the Sumatran tigers are interfertile with the Bengal tigers used to ground the term '*Panthera tigris*', the two strains belong to one species according to the BSC. But the two strains are distinct enough that each amounts to a uniquely "diagnosable cluster," so the Sumatran tiger strain would be counted a *new* species according to the PSC. Thus, a team from the American Museum of Natural History has coined the new species term '*Panthera sumatrae*' for the Sumatran tigers (Jackson 2000).

Depending on whether the PSC or the BSC prevails, scientists may conclude that the newer tigers are conspecific with the original strain, or they may conclude that they are not. Similar words apply to the Baltimore oriole, *Icterus galbula*. This bird is closely related to the Bullock's oriole, though the two strains are distinct: They look a little different and they differ genetically, but they interbreed where their habitats overlap. Whether there is one species here or two depends on whether the PSC or the BSC prevails.

Readers may wonder whether divergent rulings like those illustrated in the foregoing examples are rare. The answer is that they are not rare. The PSC is much more fine grained than the BSC and recognizes far more species. The BSC divides the birds as a whole into an established 10,000 species. The PSC would double that number to 20,000 species (see Martin 1996). The divergence is significant.

Of course, if either the PSC or the BSC is yet to be discovered to be the one true species concept, then the fact that they disagree about so many cases presents no problem for the position that earlier speakers assigned species names to various species and that scientists would later come and discover the essence of each kind. But I will argue that neither the PSC nor the BSC can be discovered to be the one true species concept. If my arguments are sound, then the many cases of divergence like the foregoing show that when earlier species terms were baptized, the essence of the species and the limits of the species would typically be no matter of discovery. Scientists might continue to battle over the matter or they might settle with one or another essence; but they will not discover that one proposed species concept is right and others wrong, so that the one concept reveals the essence of earlier named species and others do not.

Further, even when there is *no* divergence in the way scientists of different camps group species, because proponents of both the BSC *and* the PSC assign the *same* organisms to the same species, each of these accounts of what a species is still assigns a different *essence* to the relevant species. That is because each of these accounts gives a different answer as to *why* the organisms belonging to the species belong to it. Each gives a different answer as to what it takes to belong to the species. So if neither the BSC nor the PSC can be discovered to be true, there is trouble for the view that one or the other camp of scientists has discovered or will discover the essences of any earlier named species, regardless of whether or not there is divergence in the placement of particular organisms into those species. Divergence in placement merely highlights differences in essence assignments.

As I have explained, it is crucial to my position that neither the BSC nor the PSC can be discovered to be true. It is time to say more about this. On the position that I will defend, earlier use of 'species' was not precise enough to favor any one competing present-day camp. To be sure, the presence of competing views in the scientific community is not itself reason to say that an answer is up for grabs. A generation ago scientists were sharply split over whether "polywater" was an artifact. Empirical investigation showed that it was.

But despite glorious analogies (Smith 1990, p. 123, compares efforts to capture the nature of species to the search for a unified field theory in physics) and despite the many claims that this or that concept is the "correct" one,[13] the species debate seems largely to come down to a matter of personal preference that could be decided in any of a number of sensible ways. The competing proposals do have to measure up to the empirical world, and the choice between them is not completely arbitrary: As I have already suggested, it is significant that the BSC and the PSC both divide the organic world into groups that are natural and scientifically interesting. But each method for dividing the world offers different advantages, and no one method seems to trump its competition in such a way that it emerges as the one concept that properly resolves the nature of "species."

Points counted in favor of a species concept include, for example, its correspondence to systematists' vision of their field, its ease of use, and its tidiness. I will elaborate on these. Systematists' vision of their field is a factor that counts for a lot as scientists weigh different candidate species concepts. As was brought out at a celebrated meeting of ornithologists a few years ago, differences between proponents of the PSC and the BSC hinge largely on whether the discipline is seen "as a science dominated by field work or by laboratory investigation" (Martin 1996, p. 667).[14] Field workers find the BSC more manageable. It takes a laboratory specialist to distinguish species according to the PSC. The advantage of the PSC is, of course, that species names convey much more information for specialists if species are divided according to the PSC rather than according to the BSC. So whether the BSC or the PSC is preferable depends on what and whom taxonomy is for, and about this there is conflict. Some see classification as a tool for the wider scientific community and even for interested lay speakers; others do not.[15] Again, a victory for the latter would result in double-sized field guides.

A clear instance of the pull of convenience on behalf of one or another conception is manifested in the promotion of the BSC on the basis of the "considerable simplification of taxonomy" that Theodosius Dobzhansky, along with others, has found it to bring: Because the BSC accords reproductive units species status, it conveniently allows for the recognition of polytypic species (or species broken into subspecies), rather than "assigning species names to every local race" (1970, p. 356). Mayr has counted the resulting simplification of classification at the species level "the greatest benefit derived from the recognition of polytypic species" (1969, p. 38; see also 1982, p. 290; cf. Cracraft 1983, p. 165).

Tidiness also plays a prominent role in taxonomists' thinking. Nowhere is this as evident, perhaps, as in the aversion many taxonomists feel for definitions that, like the BSC, fail to place every organism into a slot, by failing to recognize asexual species. As Hull remarks, "Just as every library book must be placed on some shelf somewhere in the library, there is a strong compulsion among systematists to insist that every organism must belong to some species or other" (1987, p. 179; see also Schloegel 1999; Rosselló-Mora and Amann 2001).

There are costs and benefits to both the BSC and the PSC. Neither could ever be discovered to offer the only acceptable use for 'species', because neither *offers* the only acceptable use for 'species'. There is more than one legitimate way to settle the use of 'species', so the eventual standardization of the use of that term in one direction or another will not have been discovered to be correct.

A question arises at this point: How does what I have been saying in the past few paragraphs compare with *pluralism*, and am I endorsing pluralism about species? I have said that, depending on whether the PSC or the BSC prevails, scientists may conclude that a newly discovered strain of tigers or orioles has the essence of an earlier named species, or scientists may conclude that the newly discovered strain does not. Pluralism is a "species concept" that offers still another possibility. According to pluralists, scientists should

74

conclude that each of the strains, the newer and the original, belongs to both its own narrower species *and* to a wider species that includes both strains. So pluralists endorse a win-win solution to the problem of having to choose between attractive offers from more than one concept on the market (Kitcher 1984; 1987; Dupré 1993, pp. 44ff.; Ereshefsky 1992).[16]

The issue that I have been addressing is whether any species concept on the market has been or will be discovered to be true. I have said that the answer is no. This answer applies in view of pluralism, too. Whether or not pluralism prevails – and unfortunately its effect has been to add more competition to the market, rather than to settle the chaos – the idea that scientists might discover what makes one species concept, pluralistic or monistic, the one to adopt seems exceedingly dim. Pluralism does seem to be a sensible resolution to the species problem: the problem of characterizing the nature of species (LaPorte forthcoming). But monism also offers a sensible resolution to the species problem, and a monistic concept like the BSC has no less a claim to capturing the pre-scientific use of 'species' than the pluralistic solution according to which the species category should include groups delimited by various different criteria. Were scientists to reserve the word 'species' for groups recognized by the BSC and to call those groups recognized by the PSC something else, this would not seem to violate earlier speakers' use of 'species'. So a resolution to the species problem that favors *any* of the various parties in dispute, including the party of pluralists, seems to refine the use of 'species'.

I have opposed the position that there is only one acceptable use of 'species' and that this use accords with the use of speakers from centuries past. There is more. Suppose I am wrong. Suppose there *is* only one acceptable use of 'species' and that this use accords with the use of speakers from centuries past. Even then, there would be no particular reason to think that this one true use of 'species' will ever be *discovered* to be the one true use. If there is one acceptable use of 'species', then it would seem to be the one acceptable use because it is preferable to or more useful than alternatives. But to determine that one use is preferable to alternatives, one would have to weigh any number of highly delicate considerations that could easily fail to be properly appreciated or balanced against one another: Hence, the one acceptable use might easily fail to be adopted. Moreover, the considerations that would have to be taken into account to determine the acceptable use of 'species' would be largely non-empirical, and the task of evaluating them is not the kind of job that scientists have shown a lot of competence in tackling (as biologists often point out: Hey 2001, p. 13; Maynard Smith 2001): Whether, say, taxonomy exists to serve specialists or a larger community and whether this issue is more important than the issue of whether a species concept should cover more than just sexual organisms are not standard biological questions to be resolved by the usual empirical methods. These questions are in part biological, but they are also in large part nonbiological. Because of the difficulty of the nonbiological issues involved, even if there *is* a single correct use of 'species', there is little reason to think that scientists will discover it rather than settle on some other species concept. Two well-known systematists indicate as much:

> Realistically, the use of the term "species" will be determined as much by historical and sociological factors as by logic and biological considerations. In any case, the entities deriving their existence from different natural process [*sic*] are all valid objects of investigation. (de Queiroz and Donoghue 1988, p. 335)

It is interesting that the disputes over the use of 'species' have taken so long to die down. Darwin seemed to think that disputes like that between the PSC and the BSC would be dispelled by his theory of evolution. When evolution is acknowledged, Darwin says, "we shall at last be freed from the vain search" for the right account of how species are to be characterized (1859, p. 485):

> The endless disputes whether or not some fifty species of British brambles are true species will cease. Systematists will have only to decide (not that this will be easy) whether any form be sufficiently constant and distinct from other forms, to be capable of definition; and if definable, whether the differences be sufficiently important to deserve a specific name. (1859, p. 484)

Darwin was wrong about how soon the disputes would end. Part of what keeps systematists arguing is "whether the differences be sufficiently important to deserve a specific name." But although Darwin was wrong to suppose that the disputes would end earlier, I have argued that he was right to suggest that there is no fact of the matter that one proposal for delimiting species is the right one, and others wrong. Which conception is eventually accepted will not be a matter of discovery.

## II.2.  *Higher Taxa*

I have addressed whether scientists have discovered the essences of species, like the tiger species or the Baltimore oriole species. It is time to address higher taxa, like the mammals or the birds. In earlier chapters, I have noted that different taxonomic schools delimit such taxa differently. Cladism has become the dominant school, but it is not the only live option. Competing schools offer different answers about what it takes to be a mammal or a bird. They offer different answers about the essences of these kinds. I will focus on two schools: cladism and evolutionary taxonomy.[17]

Cladism, which finds its source in the influential work of Hennig (1966), classifies entirely on the basis of genealogy. As I have explained, each cladistic taxon, or clade, is composed of a "stem" species together with all of its descendant species. According to cladism, no member of Aves (the birds) could ever evolve into something else, because however radical the change might be, the descendants would still be descendants of the stem species of birds, and that alone would make them birds. Similarly, no nonbird, such as a moth, could ever evolve into a bird, no matter how similar to birds it became in genetic structure, ecology, and so on.

Evolutionary taxonomy, championed by Simpson, Mayr, and others (Simpson 1961; Mayr, Linsley, and Usinger 1953; Mayr 1969; Mayr and Ashlock 1991; Brummitt 2002), employs nongenealogical as well as genealogical standards in dividing the organic world. Evolutionary taxonomists, like cladists, refuse to recognize the possibility that a moth could evolve into a bird or a bird into a moth by way of convergent evolution. Evolutionary taxonomists, too, have historical standards. But unlike cladists, evolutionary taxonomists *would* recognize the possibility that a line of birds might evolve into nonbirds. If a line of hummingbirds were to lose its birdlike characteristics and take on insectlike characteristics while adapting to a new ecological niche, it could cease to be a line of birds. Though it could not become a line of insects, its members could form a *new* higher-level taxon.

Although cladism has become the dominant system, it does not seem to have been discovered to be the system that exposes the essences of kinds named before evolutionary theory; evolutionary taxonomy has as good a claim to that honor, I will argue. Hence, neither of these systems has really been discovered to be the one that exposes the essences of kinds named before evolutionary theory.

What would it take for cladism to have been discovered to be the one system that exposes the essences of kinds named before evolutionary theory? Perhaps it would have to be the best system available. On this account, earlier speakers used a term like 'bird' for a kind with an essence of the type recognized by the best available taxonomic system. I am not inclined to think that earlier speakers had anything this precise in mind, but I will suppose that they did. If they did, and if cladism is the best system available, then the rival system of evolutionary taxonomy poses little problem for essence discovery: Cladism exposes the discovered essences of the relevant variety if any system does. If, on the other hand, cladism is *not* the best system available, it does not expose the discovered essences of kinds like the bird kind. So the question is this: Is cladism the best system available? Unfortunately, there does not seem to be any fact that cladism is the best system. This is a problem for the view that cladism exposes the discovered essences of kinds named by speakers of centuries past. Further, even if cladism *is* the best system available, there remain grave problems, because of competing systems, for the position that kinds' essences are discovered to be revealed by cladism. So I will argue.

I begin by addressing whether any taxonomic system is just the best system. Of course, when the subject under discussion is genealogy, cladism has a clear edge over evolutionary taxonomy, as I have indicated in Chapter 1. But when the subject under discussion is ecology, evolutionary taxonomy enjoys a similar advantage over cladism. The crucial question here, then, is not whether cladism is the best system for representing genealogy but rather whether cladism is the best system in the context of *biology in general*. It is to this question that I will give a negative answer, as I develop a position that I suggest in Chapter 1. If the world is roughly as we think that it is, then each of these competing systems is more natural or useful for one or another purpose, but none seems most natural or useful for biology in general.

To say that no one systematic camp can claim to be best from some general biological perspective is not especially radical. Biologists themselves often acknowledge the point, although sometimes the acknowledgments appear to go too far, suggesting that classification is completely arbitrary: "Having whatever information and methods one wants, what criteria should be used to incorporate the results into a classification? This question lacks an objective answer," writes one biologist. "Only power can arbitrate genuinely basic questions of taste, and that is why there has been war in systematics" (Van Valen 1989, p. 100). Even if no one camp is best from some general biological perspective, classification is not completely arbitrary. It certainly is not the case that just any system of classification is a serious contender. No one would seriously propose a system of classification that groups organisms on the basis of color, thus grouping gray elephants and gray caterpillars together into a gray-organism group. Some systems are outside the range of candidates for becoming the standard system of classification. And those systems that are serious candidates are accepted or rejected on the basis of important considerations. Worthy costs and benefits are weighed in the selection. But each system has its own selling points, and none trumps the competition.

Evolutionary taxonomy enjoys certain advantages over cladism, which is strictly genealogical. Unlike cladism, evolutionary taxonomy enjoys the ability to reflect ecological niche, overall similarity, genetic similarity, and so on, as well as genealogy. Evolutionary taxonomists refuse to recognize some cladistic taxa that are too unnatural from the point of view of someone interested in ecology or other nongenealogical traits, and evolutionary taxonomists recognize other taxa that cladists must reject, for similar reasons.

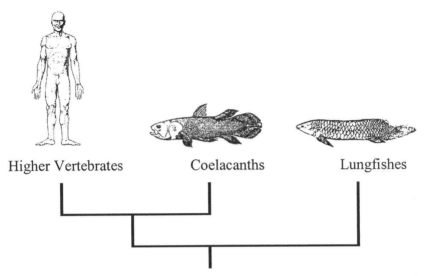

**Figure 3.1.** A cladogram showing the relationship between two lobe-finned fishes and higher vertebrates. The evolutionary taxonomist groups the fish together. The cladist groups the coelacanths with the higher vertebrates. A cladogram reveals just the order of branching, not evolutionary change, or other information. Adapted from M. Strickberger, *Evolution*, 2nd edition. Copyright 1996 by Jones and Bartlett Publishers, Sudbury, Massachusetts. www.jbpub.com. Reprinted with permission.

Thus, evolutionary taxonomists refuse to recognize the group including some but not all of the extant lobe-finned fishes, plus higher vertebrates. Instead, evolutionary taxonomists group the lobe-finned fishes together into one group and the higher vertebrates into another.

Cladists, on the other hand, must recognize a group including some but not all of the extant lobe-finned fishes, plus higher vertebrates. As unnatural as such a group seems, cladists are bound to recognize it, because some lobe-finned fishes share a closer common ancestor with us than with other lobe-finned fishes (see Figure 3.1). We evolved from lobe-finned fishes.

Cladists are unable to recognize a group that includes just the lobe-finned fishes, as *natural* as that group seems. The reason is similar. The closest common ancestor of the extant lobe-finned fishes is shared by higher vertebrates like us, so the group including just the extant lobe-finned fishes, without us, is not monophyletic.

A group like the one that includes the extant lobe-finned fishes but not the higher vertebrates is known as a *grade*: It is a group with a shared level of organization. A *clade*, by contrast, is an ancestral group and all of its descendants, some of which may have evolved so much that the organisms of the group do not have any common level of organization. By recognizing *grades*, like the lobe-finned fishes, while not recognizing unnatural-seeming *clades* like the group that includes some but not all extant lobe-finned fishes plus higher vertebrates, evolutionary taxonomy represents an element of phylogeny that cladistics misses: evolutionary change, like that which the higher vertebrates underwent after evolving the traits needed for life on land. Cladistics incorporates only the other element of phylogeny: genealogy. As a result, Mayr emphasizes, "the number of evolutionary statements and predictions that can be made for many [cladistic groups] is often quite minimal" (1976, p. 436). The group comprising a few fish plus higher vertebrates like mammals seems to have minimal predictive power, Mayr would complain. Fish have more in common with other fish than with mammals.

Some cladists concede that the number of predictions that can be made for cladistic groups is minimal, but they say that this is just the price of ignoring evolutionary change. Evolutionary change deserves study, it is admitted; it is just not part of classification (see Ridley 1986, p. 60). And whatever the cost of abandoning the ability to represent such change, the cost of retaining the ability to represent it is not negligible. Commonly noted costs are arbitrariness and confusion. The problem of arbitrariness is this: Evolutionary taxonomists sometimes recognize clades as taxa and sometimes only parts of clades, depending on the degree of evolutionary change within the clade. But decisions about how much evolutionary change is sufficient to break a clade are by no means straightforward. Thus Rosen: "Must a descendant take to the air, return to the sea, come out on land before we choose to confer high rank upon it? Or is it enough that some organ in the mouth or a part of a root system has evolved . . . ?" (1974, p. 448). The problem of confusion arises from evolutionary taxonomists' policy of including both grades and clades in their lists of taxa. As Hull records, proponents of alternative systems regularly decry "evolutionary systematists' habit of interspersing grades and clades in their classifications in such a way that no one can tell which taxa are which" (1988, p. 141). Of course, a handy system of prefixes might remedy this inconvenience.

There are advantages and liabilities for each system. Which system is best depends on what a systematist wants to represent. This suggests that there is no fact of the matter that one system is best from some general biological or scientific perspective. The different systems just reflect different information. In the same way, there can be a number of legitimate maps for a given location, each representing certain features of the represented location at the expense of others (see O'Hara 1993 for an extended and helpful development of the analogy between taxonomy and cartography). General cartographical considerations do not lead map makers to conclude that the best of maps represent subway systems, for example, or weather patterns. For some maps it is important to represent such items, while for other maps it would be distracting. Different maps of the same geographical area have different purposes and are useful in different contexts. No map can include everything of value for every purpose unless, as Lewis Carroll once suggested, a map were made to life-sized scale, in which case the region mapped could serve as its own map!

In pointing out that cladism and evolutionary taxonomy map different features of evolution, I have focused on the naturalness of each system as a means of representing the tree of life. But biologists also consider many factors other than systems' naturalness from an evolutionary or biological point of view, as they compare competing systems: Naturalness of this sort is one advantage in a system, but there are others. Systematists have been motivated to accept one system or another by the desire to provide a classification system broadly useful to workers in diverse fields (e.g., forestry, horticulture, physiology, etc.), the desire to cut speculation to a minimum and to maximize certainty in taxonomic placement, the conservative desire to avoid upsetting traditional classifications, and an abundance of other such considerations.[18] These are difficult factors to weigh against one another, which brings me to my next point.

I have expressed doubts about whether any taxonomic system can claim to be the single best system for general biological purposes, period. Now I add that even if one system *is* just best, there is no special reason to have much confidence that scientists will settle on it and hence be able to use it to reveal the discovered essences of kinds that were discussed in earlier centuries. If one system is the very best, then it would seem to be the very best on the basis of any number of highly delicate considerations that could easily fail to be fully understood or weighed properly against one another. For this reason, the best system, even if there is one, might easily fail to be adopted. Nor would any correction be needed if scientists were to arrive at some system that is less than optimal: The best system would not be needed to do the best of science. Suppose, for example, that the system that eventually becomes firmly ensconced is slightly inferior to some competing system primarily for being less user-friendly to workers in diverse fields. It will still allow for scientists to communicate first-rate results, though these results may be less widely read and appreciated by nonspecialists than they might have been with a better system. In fact, there would seem to be *nothing at all* that can be said using the best system, if there is one, that cannot be said with a system that is less than optimal: If users of a less-than-optimal system need to discuss would-be-official taxa from other systems, they can refer to them by placing the groups' traditional names or new names in scare quotes. Cladistic systematists do precisely this at present when they discuss groups that, like the "lizards," are officially recognized by noncladistic schools but not by cladism.

I have argued that no taxonomic system can claim to be the single best system for general biological purposes. I have argued a similar point previously (in §II.1) with respect to species. Readers may wonder, at this point: Although I have not explicitly committed to pluralism, does my position amount to some version of pluralism? A more worrisome question is this: Am I committed to the view that pluralism has been *discovered* to reveal the nature of higher taxa and species? If the answers to these questions are positive, then further questions arise about whether I have really escaped the position that scientists' conclusions about essences have been discovered to be true. But as I have already suggested (in §II.1), the answers to these questions are negative.

My position that no single monistic account of the nature of higher taxa or species has been discovered to be correct is shared by pluralists. But pluralists go further. Pluralists about higher-level taxa urge that various systems of classification be employed side by side, so that more than one system receives official sanction, and pluralists about species urge that 'species' be applied to groups recognized by any of a variety of species concepts. I have no objections to these suggestions, but my business here is not to recommend them, either. My business is to point out that whether scientists adopt pluralism or whether they adopt monism, what they end up with will not have been *discovered* to be the one resolution that reveals the essence of biological kinds named before the rise of modern science.

I will close this section after returning to the discussion about the term 'bird'. I have argued that although cladism is the dominant systematic school today, its answers about the essence of the birds or any other group have not been discovered to be right. Evolutionary taxonomy gives a different answer as to the essence of the birds, and it is no less qualified than cladism to give an answer.

What really happened is that cladistic scientists have refined the meaning of 'bird'. Early speakers coined the term 'bird' without any clear idea about what the essence of birdhood might be. When scientists learned more, and in particular when they discovered evolution, the vagueness became more apparent, and more bothersome. So refinement seemed in order. Some scientists opted to refine in one way, and others in another way. Cladists have won the most converts. It seems likely that cladism will remain the dominant school. Perhaps it will completely eclipse its competition, to become the only school. If so, all systematists will use 'bird' and other such terms for a group with a cladistic essence. But there will have been a change in the meaning of 'bird', not a discovery of the relevant kind's essence.


### III.   ESSENCE DISCOVERY WITHOUT THE COMPETING SCHOOLS

I have argued, in the previous section, that scientists' conclusions about the essences of kinds named by earlier speakers are not discoveries. This is so even when a word's dubbers have applied the word to all and only the members of a genuine natural kind in their environment. There are rival views about what kind of essences biological taxa have. No one of the rival views has been or will be discovered to be the correct one. In the present section, I argue that even if there were only *one* school of taxonomy and *one* species concept, it would still not be the case that scientists' conclusions about the essences of kinds named by earlier speakers are discoveries.

I begin by discussing higher taxa. A good example to introduce the present point is one from an earlier chapter, the panda. As I explained in Chapter 1, the giant panda is now considered to be a bear. But systematists disagreed about its status from the time it was discovered by Europeans in 1869 until close to the end of the twentieth century. Some systematists said it was a bear, some said it was a raccoon, and some said it belonged to a family of its own. The panda looks like a bear, but it has characteristics that set it apart from the bears with which speakers who coined 'bear' were familiar. The panda's head has a different shape, its diet is vegetarian, and its hand has an opposable "thumb." The panda does not growl or roar as other bears do but rather makes a sound that resembles the bleating of sheep. The panda also does not hibernate as other bears do. There are many differences between pandas and paradigm bears (O'Brien 1987, pp. 103–4). It is easy to see how a case like this could present problems for systematists. The panda is like the paradigm bears in some respects but not others. Is there a fact that the panda was a "bear," as speakers of 1869 used that term, or a fact that it was not a "bear"? I do not think so.

It is obvious that evolutionary taxonomists, who group in part on the basis of similarity, might easily find themselves in a situation in which there is no single right way to classify a questionable group like this. Pandas and other bears are historically related, which makes them qualified to share membership in an exclusive taxon if they are also sufficiently similar to one another. But because the pandas are intermediate between clear "bears" and clear "nonbears" in terms of similarity, evolutionary taxonomists might have a choice between placing them with what they call "bears" or not, there being no one right way to go.

It may seem that cladistics can restore principle. If cladistics has it right, all we have to do is to locate the pandas' place on the genealogical tree of life, it may seem. If they are part of the "bear" clade, then they are "bears." Otherwise, they are not. This response is suggested by the fact that the panda is now counted a bear because of its demonstrated ancestral connection with other bears.

But, interestingly, the problem for cladists is one of a kind with the problem for evolutionary taxonomists. Just as there are intermediates in terms of similarity, there are intermediate locales on the genealogical tree. There is a clade that includes all bears except the pandas, and this is embedded in another, larger clade that extends just as far as the pandas (see Figure 3.2). So the dilemma returns in a different form. The paradigm bears *without* the pandas form a taxon. The paradigm bears *with* the pandas form a taxon. There does not seem to be any fact of the matter about which of these two legitimate taxa is to be identified with the "bears." It comes down to how distinctive pandas are. O'Brien, who has demonstrated the panda's ancestry, takes the panda to be "distinctive enough to warrant placement in its own subfamily" (1987, p. 107), but he puts the panda in the bear family. Still, others might as well

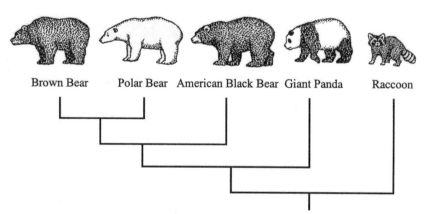

**Figure 3.2.** A cladogram revealing relationships in the line leading to the bears. Is the relatively recently discovered panda a "bear" or is it, like the raccoon, just a relative? Adapted from O'Brien (1987).

have preferred to put the panda in its own family rather than with the bears, and they would have been entitled to do so.[19]

The panda is a *living* organism found to be of an intermediate nature. Naturally, such extant intermediate organisms are not found for every taxon. Every taxon, however, has *ancestors* of an intermediate nature, so how far back a taxon extends is hardly determined by discovery. A systematist could go back only to the nearest ancestor of the group. This is the method for specifying stems that I discussed in Chapter 2. Alternatively, the systematist could go all the way back to a branch leading to some other salient group, like the closest living relative. A third method would be to go back to some inherited character (de Queiroz and Gauthier 1994, p. 29). The variety of choices is, as Hennig says, a "cause of endless and fruitless debate on the question of whether or not certain fossils should be considered 'reptiles', 'birds', 'mammals', or 'men', and when these groups evolved" (1965, p. 114).

Further, the role of the personal choice of the systematist does not end after it is decided to use some particular character or other, if that is decided upon. Different systematists choose to delimit the beginning of a clade with the appearance of different characters. For example, one systematist may take the presence of feathers to be the mark of the first bird; others may take the presence of flight to serve that function. The disagreement does not rest with a lack of empirical information (see de Queiroz 1994, pp. 503–4).

Even if just *one* character is selected to mark the start of a clade, as the appearance of feathers might be selected to mark the beginning of the birds, there may be vagueness as to when the clade begins on account of there being vagueness as to when this character arises: Modern feathers arrived after a series of more primitive structures that speakers might or might not count as feathers (see Forey 2002 for details: p. 48). Hence, different systematists might refine the use of 'bird' in different ways by drawing the line between birds and their ancestors at different points on the tree leading to modern birds, even if all of these systematists choose to mark the start of the bird clade with the appearance of feathers.

The bears and the birds are not examples of species, of course. Much the same can be said of species. Even if one particular conception of species is accepted over others, as the BSC might be accepted over the PSC, problems for determining essence remain. Conclusions about whether a given organism belongs in the same species as another will still be determined sometimes by scientists who have a choice in the matter. It can easily seem otherwise on a first glance, especially if the focus is on species at a given time rather than over time. Given that species are delimited by the BSC, it may seem that it is only a matter of discovery as to what organisms are conspecific, as one philosopher asserts:

> The intra-breeding criterion yields only *one* set of groups of living creatures, and the determination of which creatures to assign to the same group and which to different groups is no longer in any sense a matter of choice but only of discovery. (Wolfram 1989, p. 239)

Although this appears to be true at first glance, it is not true. Systematists often have a choice about whether to say there is enough introgression between two populations to consider them species. Interbreeding is not an all-or-nothing affair. This has long been known to biologists, and has often been commented upon: "[T]he infertility of species when crossed graduates away so insensibly that the two most experienced observers who ever lived have come to diametrically opposite results when experimentising on the same forms," Darwin writes (1975, p. 165). Essentially the same point has been made by many contemporary authors (e.g., Dobzhansky 1970, p. 359; Hey 2001, pp. 20–1; Johnson 1970, p. 231; Mayr 1982, p. 282; Mishler and Donoghue 1982, p. 495; Templeton 1992, p. 165).

Therefore even if we restrict our attention to the BSC, it will often be questionable whether an older species term is to be extended to a newly discovered population, or whether a new species term is needed. There often seem to be no facts to determine the matter. Biologists simply have to choose whether to apply 'species' to the larger or to the smaller group.

Much the same can be said about other species concepts. The other concept most discussed above is the PSC, which defines species as "smallest diagnosable clusters." But, as Mishler and Brandon note, "Some judgment of the significance of discontinuities is needed" (1987, p. 408).

The ecological concept, endorsed by Van Valen (1976) and Andersson (1990), characterizes a species as a lineage with a unique adaptive zone, or ecological niche. But again, adaptive zones vary gradually, and are very difficult to specify, so that the systematist will find herself splitting here and bonding there where she might have done otherwise (see Rosen 1974, p. 448).

Also popular is the evolutionary concept, originally proposed by Simpson (1961, pp. 152ff.) and slightly modified by Wiley (1978). According to the latter, "a species is a single lineage . . . which maintains its identity from other such lineages and which has its own evolutionary tendencies and historical fate" (p. 18). Again, the taxonomist is hardly left with cut-and-dried species waiting to be discovered. Just when a lineage can be said to have its own "evolutionary tendencies" and "historical fate" calls for judgment.

It seems hard to believe that there should be *any* concept that obviates the need for systematists' discretion. On the contrary, as Simpson has observed, there are bound to "remain numerous doubtful cases where decision depends on the personal judgment of each practitioner of the art of classification. To insist on an absolute objective criterion would be to deny the facts of life, especially the inescapable fact of evolution" (1961, p. 152).

The mapping of previously coined terms to the tree of life after empirical disruption is no straightforward affair, even given a single systematic school or species concept. Even given a school or species concept, mapping must often be achieved by stipulation, not discovery.

## IV. VERNACULAR USE

Sections (I)–(III) address how scientists continue to use terms as empirical knowledge grows. I have looked at how evolutionary findings bring about changes in scientists' use of terms. The terms whose use is most clearly affected by such findings are scientific terms, such as '*Cavia porcellus*' and 'Rodentia'. Scientists' use of vernacular terms like 'guinea pig' and 'rodent' also seems to be affected by evolutionary findings in the manner that I have described.

In accordance with the usual position that scientists determine vernacular use, I have supposed that ordinary speakers' use of vernacular terms like 'guinea pig' and 'rodent' is affected in much the same way as scientists' use of these terms and their correlates. Whether ordinary speakers' use of kind terms is affected by evolutionary findings in the same way that scientists' use of kind terms is affected by such findings is in need of defense, however. Indeed, whether ordinary speakers' use of kind terms is affected *at all* by evolutionary findings is in need of defense. I provide the needed defense in the present section.

### IV.1. *Evolutionary Findings Are Relevant to the Vernacular*

According to some philosophers,[20] findings like those suggesting that guinea pigs evolved separately from mice and rats are just irrelevant to the use of vernacular terms. Dupré captures the sentiment when he says of a similar example, "There may be a case for such linguistic revisions in professional systematics, but there can be no possible case for holding ordinary language hostage to such quirks of evolutionary history" (Dupré 1993, p. 33).

If Dupré is correct here, then contrary to what I have assumed, ordinary speakers' use of terms like 'guinea pig', and perhaps 'rodent', is unaffected by scientific findings. But Dupré is not correct here. As I have already argued (see Chapter 1, §II.5), vernacular terms are not insulated from the changes in use that attend their scientific correlates. Influence from the direction of evolutionary theory on the proper use of vernacular terms (including ones introduced by science, like 'rodent', perhaps) is powerful and widespread. As the scientific journal *Nature* published the foregoing findings about the rodents, so did lay publications, like *The New York Times* (Angier 1996). And the *Times*'s front-page announcement was not about scientific terminology, or technical uses for 'rodent': It was simply about rodents and guinea pigs. "People may think they recognize a rodent when they see it scurry by in the park," the paper announced, cautioning that recent scientific work has undermined the importance of popular operational criteria.

Another example mentioned above is the use of 'dinosaur' and 'bird'. A few years ago a professional taxonomist described to me his bewilderment upon first reading a textbook that called birds "dinosaurs."[21] He did not know what this talk about "avian dinosaurs" referred to, until he figured out that the book was talking about birds. Such talk is now filtering into the mainstream media. In the daily news, scientists are quoted saying things like "It is now impossible for any credible person to claim that birds are not theropod dinosaurs."[22] This is how scientific use is passed on to the laity. Children's literature and lectures pass the lesson to new generations. Children at one educational program are told, "When you eat turkey or when you eat Chicken McNuggets, you're really eating dinosaur."[23]

My point is not, of course, that the conclusion "Birds are dinosaurs" marks a discovery about what we have all along called "birds" and "dinosaurs." Scientists could as well have concluded, "Birds are *not* dinosaurs," and some scientists would like revision to go that way. These scientists, for example Patterson (1993), say that the "dinosaurs" do not constitute a complete clade. Patterson does not extend the use of 'dinosaur' to make it apply to birds, which are not traditionally recognized as dinosaurs. Interestingly enough, he does extend the use of '*bird*' to make it apply to dinosaurs, which are not traditionally recognized as birds. For Patterson birds are not dinosaurs; it is the other way around. "Perhaps surprisingly, dinosaurs turn out to be kinds of bird" (1993, p. 518), he insists.

Patterson's gratuitous application of 'bird' to the likes of *Tyrannosaurus* and *Stegosaurus* underscores my claim that terms can be adjusted in a variety of ways in the face of disruption. It is also a good illustration of my point from the previous section that even if cladism were the only school, the boundaries of a kind like the dinosaur or bird kind would not be discovered by cladists, because there are different cladistic means for choosing a clade's origin, or stem. Patterson chooses a different origin for the birds than most cladists.

So scientists' conclusion "Birds are dinosaurs" was not discovered to be true. But it is becoming generally accepted.[24] Once scientists draw a conclusion like this, lay use is at least likely to follow. Lay use is certainly not immune to this sort of revision. The fact that 'dinosaur' is now a vernacular term does not undermine my position that empirical discoveries are relevant to the term's use.

## IV.2.  *Multiple Uses and the Vernacular*

Vernacular use is often revised in light of science. But even though science is relevant to the reference of vernacular terms, there may be slight differences between the vernacular use of a term and a corresponding scientific use. In Chapter 1, I maintained only that vernacular use *approximates* scientific use. There may be enough of a difference that pressure from empirical findings could sometimes result in the severing of scientists' use of a term from vernacular use. This would hardly seem to be a general rule, but it may be an occasional outcome. 'Dinosaur' might, arguably, have been settled by a split between lay and scientific use.

I have pointed out that scientists, or at least the majority of scientists, have responded to the news that birds descended from dinosaurs by rejecting 'There are no more dinosaurs'. If it is a possible outcome of empirical disruption that scientists' use of a term is severed from vernacular speakers' use, then ordinary speakers might respond to such disruption by saying something like "There are no more dinosaurs. Zoologists say otherwise, but they use 'dinosaur' in a special, technical sense, not as you and I use it." I have already emphasized that it is not at all clear that lay speakers would respond this way.[25] But suppose lay speakers might respond this way. Even if they might, the basic picture I have offered of the continual refinement of vagueness seems to remain sturdy. Scientists refine the use of 'dinosaur', and so would lay speakers. It is just that these two groups of speakers could refine the use of 'dinosaur' differently.

By continuing to deny that there are "dinosaurs," lay speakers would refine; they could have said that there are "dinosaurs" with as much justification, given the surprising news that birds are the extant progeny of dinosaurs. After all, when lay speakers and scientists used to say things like "There are no more dinosaurs," they supposed that this was so because the dinosaurs had no living descendants. The possibility that dinosaurs have descendants but that these descendants have evolved into spectacularly different forms was not really considered. When the legacy of dinosaurs came to light, it was not clear what to say about the truth value of the claim "There are no more dinosaurs."

Other responses to the disruption could occur: Lay speakers could respond, for example, by giving 'dinosaur' a span of senses in the vernacular, so that the original vagueness is no longer present, but there are many refined newer senses, or different uses in different contexts.[26] Occasionally, too, there might be no vagueness refinement at all: The full original vagueness in a term, exposed by empirical research, may remain in all of the various contexts of use. Refinement does, nevertheless, seem to be the rule. Scientists are self-conscious about redefining terms for the sake of precision, as I have indicated. Educated lay speakers often follow, because after all they learn about organisms from scientists. And even if lay speakers do not follow scientists, they will want to refine the use of terms at least in particular contexts. Earlier speakers' claim "There are no more dinosaurs" was of unclear truth value, because birds are a questionable case. Because there is usually little point in making claims that we know are of unclear truth value, speakers can be expected to refine. Thus if an informed speaker says, "There are no more dinosaurs," she discounts birds. If she says, "There are still dinosaurs," she counts them. Either way, there is refinement.[27]

## V. CONCLUSION

It is time to take stock, draw conclusions, and then introduce the link between these conclusions and upcoming chapters. I have argued against the common claim that scientists' conclusions about kinds' essences are generally discovered to be true. I have also argued that when scientists lead lay speakers to stop calling a particular group of organisms by a kind term that these speakers have habitually applied to the group, or to begin calling a particular group of organisms by a kind term that these speakers have habitually refrained from applying to the group, scientists have not discovered that the new use is right and the old use wrong. 'Rodent', for example, was not discovered not to apply to guinea pigs, though scientists have concluded that the guinea pig is not a rodent.

My own account of how terms are revised as science advances is that vagueness and confused suppositions underlying the use of the terms are removed after empirical light is shed on items to which the terms have been applied. Vagueness can be dispelled in any of several ways of modifying a term's meaning, no one of which can claim to preserve the exact original meaning of the term.[28]

Conclusions that mark empirical advancement, like "The guinea pig is not a rodent," often, therefore, seem to be established in part by meaning change. This kind of change in both theory and meaning is ubiquitous in science and other intellectual pursuits, like philosophy. A proper understanding of what goes on in such cases of change is therefore crucial for an understanding of conceptual advancement. I address important issues surrounding conceptual advancement at some length in Chapters 5 and 6.

In Chapter 5, I relate the considerations of the present chapter and the one following it to the philosophy of Thomas Kuhn. According to Kuhn, the meanings of theoretical terms change through scientific upheaval. Putnam insists that this is not the case, using Kripke's theory of reference to argue that meanings do not change through scientific change. If the arguments of the present chapter are sound, however, Kuhn is right. Meanings *do* change. Questions therefore arise, as Kuhn makes clear, about the extent to which scientific changes in position can amount to genuine progress, as opposed to a mere change of terminology.

In Chapter 6, I address a question that has, no doubt, occurred to many readers. The question is whether theory change and meaning change *inevitably* attend one another for the simple reason that there is no *difference* between theory change and meaning change. My own investigations above into conceptual change certainly might appear to lend strength to the claim that theory change and meaning change cannot be distinguished.

Before addressing these issues about conceptual change, I turn my attention briefly from biological kinds to chemical kinds. I hope to show that the problems for discovery with respect to the essences of biological kinds that I have presented in the present chapter apply more generally. Paradigmatic natural-kind terms are taken from chemistry, as well as biology. Therefore a brief examination of kinds from chemistry is well worth pursuing before I move on.

# 4

## Chemical Kind Term Reference
## and the Discovery of Essence

According to the familiar view of scientific inquiry that I have been examining, scientists draw conclusions about the nature of kinds that were named before much science was known. Birds are dinosaurs, scientists have concluded, and water is $H_2O$. These conclusions are supposed to be discoveries about the essences of the bird kind and the water kind. The words 'bird', 'dinosaur', and 'water' are not supposed to have changed in meaning.

I have argued that this account of the conceptual change in question is not right. I have argued that biologists' conclusions, or future conclusions, about the essences of kinds recognized before current systematic theory are in general not discovered to be true. I have focused on kinds from biology, but similar words apply to kinds from chemistry, the other major source of natural-kind terms. My aim in this chapter is to show that.

Much of the motivation for supposing that scientists' conclusions about natural kinds' structures are discoveries about essences comes from Hilary Putnam's famous Twin Earth thought experiment, which concerns a chemical kind: water (Putnam 1975e, pp. 223–7). Twin Earth is a distant planet just like ours except that the waterlike substance on that planet has a long, complicated chemical composition that is abbreviated 'XYZ'. XYZ looks like water on Earth, tastes like it, comes down in the form of rain, fills rivers, lakes, and oceans, and so on. Yet Putnam says that XYZ does not belong in the extension of *our* word 'water', even if inhabitants of Twin Earth call it "water." If we were to travel to Twin Earth, we would probably initially *think* that XYZ is water, because it is so similar to water superficially, but after we discovered its underlying composition we would correct ourselves and refuse to call it "water." We would say that it lacks the microstructural composition of what we call "water." We reserve 'water' for substances sharing the microstructural essence of samples called "water" on *Earth*. Those samples are $H_2O$. No other substance can be water.

So being $H_2O$ is supposed to be a *necessary* condition for being called "water." It is also supposed to be a *sufficient* condition. If we were to find a substance quite different in appearance, texture, and so on from the stuff with which we are familiar, and which we call "water," we would start calling the new stuff "water" too, if we learned that it is $H_2O$ (see especially Kripke 1980, pp. 128–9).

Because being $H_2O$ is supposed to be both necessary and sufficient for being water, the essence of water is supposed to be $H_2O$. I do not think that the essence of water is $H_2O$. On the contrary, I think that we might conclude, if confronted by XYZ, "XYZ is water." We might also conclude, if confronted with $H_2O$ that has unusual characteristics, "Some $H_2O$ is not water."

I do not argue that if we were confronted by XYZ then we would *have* to conclude, "XYZ is water," on pain of error. It is not that we would be right to affirm 'XYZ is water' and wrong to reject that statement, as a well-known *minority* response to Putnam would have it: According to this minority response (e.g., Mellor 1991), XYZ would simply be water. I argue instead that XYZ is neither clearly in nor clearly out of the extension of the vague word 'water'.

Vagueness tends to be ironed out over time, so a definite answer as to whether 'XYZ is water' is true would be likely to emerge if we were to find any XYZ. It is just that the conclusion would refine the meaning of 'water'. Contrary to Kripke and Putnam, 'water' would change its meaning.

On the view that I put forward, then, the descriptions associated by speakers with a kind do *not* completely determine the extension of a word: Being clear, tasteless, and wet is not straightforwardly necessary and sufficient for a substance's being properly called "water." On the contrary, it seems that the microstructure of samples plays some role in the reference of 'water', just as Kripke and Putnam say. But the familiar qualities of water also play some role in the reference of 'water', contrary to what Kripke and Putnam say. When the familiar qualities and the underlying structure come apart, there is no clear answer as to what we should say: Such a split exposes vagueness.

I will argue my case as follows. In section (I) I argue that when a substance has the right observable properties but the wrong microstructure, it is a vague case. In section (II) I argue that when a substance has the wrong observable properties but the right microstructure, it is a vague case. In section (III) I set aside my arguments about observable properties from sections (I) and (II) and suppose that essences are straightforwardly microstructural. There I argue that it seems wrong to say that scientists have discovered the truth of the sentence 'Water is identical to $H_2O$ *rather than some other microstructure*'. In section (IV) I wrap up the chapter by remarking on issues that remain to be addressed.

I.  SAME PROPERTIES, DIFFERENT MICROSTRUCTURE

The lesson about 'water' and XYZ is supposed to apply broadly, of course, to chemical kinds in general. I will return to the example of water shortly, but for the next couple of sections I will look at minerals. Many of the natural-kind terms from chemistry that are commonly cited in the literature name minerals recognized since antiquity for their beauty: 'diamond' (Donnellan 1983, p. 86; Margalit 1979, p. 29; Sterelny 1994, p. 10), 'emerald' (Teller 1977, p. 174), 'jade' (Putnam 1975e, p. 241; Boër 1985, pp. 114ff.), and so on. We have a fair amount of historical data on these sorts of terms, as their referents' charm captured the attention of early speakers who named them, discussed them in their literature, and so on.

## I.1.  *A Historical Twin Earth*

One mineral term with a particularly interesting history is 'jade'. This term has received fairly frequent discussion in the literature because Putnam (1975e, p. 241) uses it as an example to illustrate what happens when a term is unknowingly applied by speakers to two different chemical kinds before chemical composition is understood. Putnam clarifies that this example is quite different from that in the story of Twin Earth. In the story of Twin Earth, 'water' is applied by speakers on Earth to $H_2O$ alone for centuries, and *then* these speakers find XYZ. Putnam says the speakers from Earth will conclude that XYZ is not designated by 'water'. The story of jade, Putnam says, is very different. 'Jade' was applied by speakers to *two different* chemicals for centuries, he says. No new material like XYZ was introduced after the term had an established application to its two chemicals. Putnam says that because no new substance was introduced, the term should go on including two chemicals in its extension.

The philosophical community has up until this point taken Putnam's empirical information as data. Philosophers have sometimes talked about whether Putnam can handle the case of jade just as he says he can or whether some other treatment is required. But no one has questioned Putnam's empirical information.

My own research indicates that Putnam's empirical information is not right. If I am correct, the real history of 'jade' presents a much more difficult case for Putnam than he realizes. Indeed, if I am right, 'jade' presents a historical *counterexample* to Putnam's Twin Earth lesson. The story of 'jade' is a lot like the story of Twin Earth. The main difference, if my information is right, is that *real* speakers did not evince dispositions of the sort that Putnam

94

is committed to saying that they would. In the historical case of jade on Earth, which resembles Putnam's imaginary case on Twin Earth, speakers encountered a new substance with properties similar to those of a formerly recognized substance but with a completely different microstructure, and speakers responded by applying a term for the formerly recognized substance to the new substance.

The Chinese relationship to jade, or "yü," as they call it, is more interesting than Westerners' relationship to the material, so I will devote most of my attention to the Chinese term 'yü'. For the Chinese, jade has enjoyed something like the status gold has enjoyed in the West. The Chinese consider jade to be the most precious of material substances, more precious than gold, as a Chinese saying indicates: "One can put a price on gold, but jade is priceless" (Ward and Ward 1996, pp. 9–10). In contests of skill in ancient China the victor received a scepter of jade, not gold; the second-place competitor received a scepter of gold (Gump 1962, p. 15). The Chinese once used jade as a coin (Sakikawa 1968, p. 34). Its value in jewelry is comparable to that of gold and diamonds in the West. Most salient of all is jade's value in masterful carvings. Westerners express excellence by comparing a person, character, performance, or other item with gold or silver, as in "heart of gold," or "silver-tongued." The Chinese express excellence by comparing a worthy item to fine jade.

Many jade-like materials are not jade. As one might expect, authenticity is of the utmost importance to the Chinese. Jade impostors are most unwelcome (Kraft 1947, pp. 13, 19).[1] Both 'gold' and 'yü' have been used to designate, at least until relatively recently, only substances with a particular chemical composition. Gold is an element, so it has turned out to have a simple chemical formula: Au. The jade used since ancient times by the Chinese is nephrite. Nephrite has the chemical formula $Ca_2(Mg,Fe)_5Si_8O_{22}(OH)_2$. Kripke and Putnam are committed to saying that if a totally new substance, just like these chemicals in appearance, were to be discovered, the new substance would not belong to the extension of 'gold' or 'yü'. This new substance would be like XYZ: Speakers might at first mistake it for the familiar kind it resembles, but they would correct themselves when they learned its chemical composition.

History offers a different lesson.[2] Amazingly, jade met its XYZ. After thousands of years during which the Chinese had worked with nephrite, a very similar stone with a totally unrelated composition, $NaAl(SiO_3)_2$, made its way for the first time to China. A large shipment of the stone was brought from Burma near the end of the eighteenth century. Less than a hundred years later, in 1863, the French scientist Alexis Damour determined that this new substance has a totally different chemical composition from that of nephrite. He coined the term 'jadeite' for the Burmese stone.

But the Chinese, who had carefully worked nephrite by hand for generations, could tell by its feel that this was a different material (Gump 1962, pp. 181–2). They called it "new jade" to distinguish it from nephrite jade (Keverne 1991, p. 23). They also called it "Yün-nan jade," because it entered China through the province of Yün-nan. The most popular name for it was probably 'kingfisher jade', which was applied because the color of jadeite can resemble that of the kingfisher's feathers (see esp. Hansford 1948, p. 16).

The question that faced the Chinese is whether this new jade, or kingfisher jade, was "*true* jade." Nephrite was unquestionably "true jade" and had always been recognized as such. Jadeite was not so clear a case. The fact that 'jade' ('yü') was used in compound expressions like 'new jade' and 'kingfisher jade' does not, of course, indicate that it was "false jade." This is especially clear because similar expressions refer to nephrite. If jadeite came to be called "new jade," nephrite came to be called "old jade." 'Kingfisher jade' suggests a common color of jadeite, and similar expressions are given to nephrite of various colors: Different shades of white nephrite are called, for example, "mutton fat jade" and "camphor jade." Indeed, 'kingfisher jade' is a term that had itself been applied to a green shade of nephrite from Turkestan before the term was applied to jadeite.[3]

Would speakers accept the new kingfisher jade as true jade? Yes, they would. They have. Significantly, the Chinese have come to accept jadeite as true jade. 'Yü' and 'jade' both apply clearly to jadeite as well as to nephrite. True jade includes and is limited to nephrite and jadeite. Jadeite has now even surpassed nephrite in use.

The acceptance of jadeite as the real thing is remarkable, in view of the venerable status of yü in Chinese culture. Authorities have marveled at the phenomenon. Thus, Ward and Ward (1996, p. 24):

> For the past two hundred years (and disregarding the 5,000 years that preceded them) jadeite has been the preeminent stone and gem within China. It seems that no one objected to the culture's central substance being supplanted by a totally different material. Perhaps calling both *yü* eased the transition.

## I.2.    *Open Texture Refined*

I do not think it would have been straightforwardly correct to call jadeite "true jade" or "true yü" when it was first discovered. Nor do I think it would have been straightforwardly *in*correct to call jadeite "true jade" or "true yü" when it was first discovered. I think that the matter had not been considered, so was simply unclear. Here was a new material, very much like the old one in properties that made speakers value the older material; but it had a different composition. Because these two means of distinguishing "true jade" came apart, the matter was vague. Hidden vagueness in a word's application that is later exposed like this with more information is known as *open texture*.[4]

There has been a subtle change of meaning in the term 'jade', or 'yü', as vagueness has been refined away. Now the term applies clearly to jadeite, though jadeite was an unclear case before. One authority puts the matter aptly:

> Present-day, knowledgeable Chinese (and Westerners, too) now acknowledge both nephrite and jadeite as true jade. So it seems that, after centuries of people's not knowing or not caring, a decided drift has taken place toward [counting] nephrite and jadeite as sole inheritors of the name *yu* or *jade*. (Desautels 1986, p. 2)

One might suppose, on the contrary, that jadeite always belonged in the extension of 'yü'. 'Yü' always meant *nephrite or jadeite*, or perhaps *any stone enough like nephrite in observable properties*. That is one candidate for being the moral of the foregoing story of 'jade'. It is not a promising candidate. If it were the right moral to draw, then speakers would have been straightforwardly mistaken had they said, when first presented with jadeite, "This is not true jade." But it is hardly clear that speakers would have been just mistaken to say this, in view of the historical use of the term 'jade' for another chemical. The new stone could have been given a new name just as easily as not, by virtue of its totally different composition. The newly recognized and newly named stone could then have taken a place *beside* "jade," or nephrite, in stone carving and related activities. It would have been counted "jade's younger rival."

It does not seem very plausible to suppose that before jadeite's discovery it clearly belonged in the extension of 'yü'. Also unpromising is the contrary position that jadeite clearly *failed* to belong in the extension of 'yü', so that when speakers decided to call it "yü" they merely changed the meaning of 'yü'. That is another moral some might be tempted to draw, and if it is the right moral to draw then Putnam's Twin Earth lesson can be salvaged intact. But the claim that that is the right moral to draw seems ill supported and motivated only by a desire to save a theory.

Putnam and Kripke prompt intuitions about the proper use of a term in counterfactual scenarios by asking *what we would say* were we presented with this or that scenario. This procedure seems reasonable. Speakers' dispositions indicate the proper use of a term. And Chinese speakers have indicated by their actions *what they would say* were they presented with a new substance like what they had called "jade" except in its microstructure. These speakers have been presented with such a substance, and they have displayed a strong disposition to count it "jade." Therefore, speakers' dispositions can hardly be said to make a case for the position that jadeite would clearly have failed to belong in the extension of 'jade'.

Friends of the intuitions of Kripke and Putnam who are convinced that microstructure trumps superficial properties in determining reference might suggest that speakers' dispositions about 'jade' be disregarded. But friends of the intuitions of Kripke and Putnam are not entitled to disregard speakers' dispositions, at least not in general. Speakers' dispositions are what motivate in the first place the conviction that microstructure trumps superficial properties in reference. Nor does it seem principled to disregard speakers' dispositions just in selected cases like this one, in which those dispositions point in an unwanted direction.

One might hope that when jadeite was at first clearly understood and appreciated, it was unequivocally counted as "non-jade," and that after that there was a clear decision, maybe a ruling, to change the meaning of 'jade'. But there is scant evidence for such a story. To be sure, there is some evidence that before jadeite was well known and well understood, it was ill regarded: Some evidence suggests that the Chinese suspected at first that jadeite was a cheap imitation. But after the Chinese had worked a lot of the material and considered its properties, their esteem for the material improved dramatically. There is no reason to suppose that speakers' considered view of the new material is that it was simply "non-jade." Rather, it seems that after speakers had carefully examined the new substance and considered how closely it resembled nephrite in so many respects, and also how competitive it would be with nephrite on the market because of its rarity and so on, speakers had a strong disposition to count the new material "true jade."[5]

Not that the disposition to count the material "true jade" would not have been balanced by a competing disposition to count the material something else. But eventually, the disposition to resist counting jadeite "true jade" would fade. Then there would remain only a disposition to count jadeite "true jade," and jadeite would clearly belong in the extension of the expression 'true jade'.

The reason speakers had a disposition to count jadeite "true jade" after consideration is no doubt that its resemblance to nephrite is quite thorough, so that for practical purposes, there is little reason to discriminate between the two substances. The Chinese technique of jade carving is the same for both nephrite and jadeite (Hansford 1948, p. 15). Both materials are admirably tough, which is to say that it takes great force to damage them by crushing or smashing. Nephrite is the world's toughest stone (Gump 1962, p. 24). Both nephrite and jadeite are also very hard. A knife will carve a furrow in soapstone or other imitations, but not in jade of either type (Wills 1964, pp. 13–14; Spencer 1936, p. 213). Jadeite is slightly harder than nephrite, but this difference and others between the two minerals can be overlooked in ordinary situations. Because, as one author writes, "the resemblances between the two are many and the distinctions limited to matters of hardness, specific gravity, and x-ray analysis – not commonly considered in setting prices of jewels for commercial consumption" (Sakikawa 1968, p. 35), their similar treatment seems natural. Other authorities agree. Writes Wills (1964, p. 14):

> In the main, it is correct to say that the differences between jadeite and nephrite are of greater interest to archaeologists, mineralogists and geologists than they are to collectors. They do not greatly affect the working of either stone, and the wide range of colors of both will be found to overlap so that no rigid rules can be made for their certain identification. The wide use of the word jade is therefore not unreasonable. . . . Finally, it may be added that the commercial value of a piece is not normally affected by the fact that it is made from one or the other: both are jade to the dealer and the collector.

### I.3.  *Jade in the West*

The history of the term 'jade' in the West is more complicated. Westerners of centuries past had much less exposure to jade and much less of a taste for it. Marco Polo is thought to have described nephrite jade when he visited China in the thirteenth century. Later, Westerners found jadeite in America, where Native Americans revered the material much as the Chinese revered their jade. Still later, nephrite would be shipped to Europe from China. Then the chemical composition of jade would be unlocked.

After Damour's discovery that jadeite and nephrite are two different chemicals, there has been some inclination in the West to restrict the proper use of 'jade' to nephrite. In the year 1911, some forty-three years after Damour's discovery that jadeite and nephrite are two different chemicals, a standard general reference source had this to say about jade:

> **Jade**, a name commonly applied to certain ornamental stones, mostly of a green colour, belonging to at least two distinct species, one termed nephrite and the other jadeite. Whilst the term jade is popularly used in this sense, *it is now usually restricted by mineralogists to nephrite.* (Rudler 1911, p. 122; my emphasis)

Today 'jade', like 'yü', designates both nephrite and jadeite. Again, there seems to have been refinement in the general use of the term. The inclination to restrict 'jade' to nephrite yielded to the inclination to use it also for jadeite. Some speakers inclined to count just nephrite "true jade" (Rudler 1911, p. 123), but other speakers inclined to grant jadeite the honor. After some lack of clarity about the proper general use of this term, the inclination to count jadeite "true jade" came to dominate throughout the language. This uniformity cleared up vagueness and confusion in much discourse. The benefits for communication provided motivation for self-conscious speakers to keep to the newly refined, general use. As one authority in 1936 advises, "confusion is introduced when the term jade is limited to nephrite" (Spencer 1936, p. 212).

### I.4. *The Moral of Jade*

History seems to belie Putnam's intuitions about Twin Earth. The proper extension of a term is determined not only by microstructure but also by relatively easily observed properties like color, texture, weight, hardness, taste, and so on. Usually pure specimens with the right microstructure have the right properties and vice-versa. I have suggested, however, that if a substance with the right properties and the wrong microstructure is found, it is a vague case.

I have argued the foregoing claim by appealing to 'jade', but the lesson would seem to apply more broadly to 'gold', 'water', and so on. If we were to find XYZ, we would have found a substance that is a borderline member of the extension of 'water'. We might call XYZ "water," contrary to Putnam. Then again, we might not: We could go either way.

### II.  SAME MICROSTRUCTURE, DIFFERENT PROPERTIES

I have addressed the question of whether we would ever say that something without the right microstructure is properly called by a term like 'jade' or 'water'. It is time to address the question of whether we would ever say that something *with* the right microstructure is *not* properly called by a term. Kripke (1980, pp. 128–9) takes chemical composition to be sufficient for the application of 'water'. He suggests that if a new substance were found that had the chemical composition $H_2O$ but that had observable properties that were very different from those that we expect water to have, speakers would still call it "water." Again, history suggests otherwise.

After the discovery of the chemical composition of charcoal, scientists hardly expected the composition of this humble substance to be shared by any impressive material. But chemists were surprised by what they found after investigating diamond (Dietrich and Skinner 1990, p. 1). Amazingly, diamond's chemical composition was found to be *exactly the same* as that of *charcoal*. Still, we do not say that diamonds are charcoal or vice-versa, or that these are two varieties of a single species. Rather, we say that something besides chemical structure matters to what counts as a member of the substance charcoal.

Just as speakers have not called diamond "charcoal," even though it has the very same chemical composition, speakers might not call a substance "water" even if it has the very same chemical composition, $H_2O$.[6] $H_2O$ that is unlike our familiar water might be said to be "non-water." That is not to say that it would *clearly* be non-water. Again, it would be a vague case, and speakers might go either way with the term 'water'.

In the same way that $H_2O$ with different properties would not clearly be non-water, carbon with different properties was not clearly "non-charcoal" when diamond's composition was revealed. When the scientist Smithson Tennant first learned of the chemical affinity between diamond and charcoal, it was unclear whether he should say that diamonds are "charcoal." Speakers have refined the use of 'charcoal' to give a clear answer: No, diamond is not charcoal, nor does it consist of charcoal. Both substances consist of carbon. But speakers might have concluded, "Diamond is charcoal in an unusual state," or "Diamond consists of charcoal." Tennant himself inclined to this conclusion. The conclusion he draws from his experiments is that diamond "consists entirely of charcoal, differing from the *usual* state of that substance only by its crystallized form" (Tennant 1797, p. 124; my emphasis).

Many other examples suggest a similar point. 'Ruby' is a venerable term that has long been used for a mineral that is composed of the chemical compound $Al_2O_3$. Earlier speakers did not know this composition; they managed to identify the stone by its superficial properties, one of which was its red color. The original meaning of 'ruby' was *red*.

Eventually scientists revealed the composition of rubies. At that time, late in the eighteenth century, mineralogists were surprised to learn that this mineral comes in *blue* and other colors, as well as red (Hughes 1990, pp. 1ff., 15). A few impurities cause color variation. One might expect, then, that 'ruby' would be applied to the blue stones. But it was not. When it was eventually realized that the things called "ruby," all of them red, were specimens of a mineral that comes in many colors, people nonetheless continued to reserve 'ruby' for the red specimens of that mineral.

Speakers were able to continue to call only red stones "rubies" not by ignoring science but rather by interpreting 'ruby' as a name for a mineral *variety* instead of an entire species. The reader might suspect that varietal names for minerals are unusual, so that 'ruby' is a special case, but that is not so. Myriad kinds distinguished by superficial features have similarly been named. Terms that name kinds demarcated by color or other such characteristics are common. These are recognized by both specialists and lay speakers (see, e.g., Clark 1993).

Nor would it be right to say that "ruby" was *discovered* to be a variety of the mineral that composes it, corundum. On the contrary, whether "ruby" can be blue was unclear when speakers first learned that there are other colors of the relevant mineral. Speakers could as easily have started applying 'ruby' to the other colors of that mineral as not. Other historical cases indicate this. 'Topaz', for example, was used to refer to a mineral that, unknown to users from centuries past, is composed of the chemical compound $Al_2SiO_4(F,OH)_2$. Topaz was picked out by its superficial properties. One of those properties was its brilliant *yellow color*. The original meaning of 'topaz' was *fire*. Although only yellow stones were originally called "topaz," speakers learned, after some technical development, that some known *blue* minerals have basically the same microstructural composition as what they had been calling "topaz." Among other similarities, the blue minerals' chemical composition is also $Al_2SiO_4(F,OH)_2$. In this case speakers responded differently than in the ruby case. Speakers concluded that blue specimens of the mineral *are* topaz.

That 'topaz' refers to all of one mineral and 'ruby' to only the red of another seems to represent decision, not discovery. If such conclusions *were* discoveries, then, as many suppose, we would be in a position to correct past speakers who differed with us about whether this or that thing belongs to the extension of 'ruby' or 'topaz'. In fact, retrospect gives us no such authority.

It might be argued that terms like 'ruby' and 'topaz' are not natural-kind terms, so that the claims of Kripke and Putnam cannot be expected to apply to them. But this is not right. 'Topaz' is a natural-kind term. 'Ruby' is not a natural-kind term. Although rubies are all stones of one kind of mineral, the ruby kind is color restricted. Still, this does not threaten the significance of the foregoing. What it shows is that it may be unclear, before the rise of science, whether a substance term refers to a natural kind, or to a color-restricted artificial kind. This is similar to the indeterminacy that I have noted attends certain biological kind terms. 'Ruby' could have turned out, like 'topaz', to refer to all colors of its composing mineral. Then it would have referred to the mineral itself, which is a natural kind, rather than to the red variety of the mineral, which is not a natural kind.[7]

## III.   RELATED MICROSTRUCTURES

I have argued that vernacular terms for substances like water or minerals are vague in such a way that it is not clear whether a new substance with the familiar superficial features but a different microstructure, or a new substance with unfamiliar superficial features but the same microstructure, would belong in the extension of the term. In the present section I argue that even if we put aside these issues and assume that our vernacular terms definitely refer to microstructural natural kinds, still there will be trouble for the view that scientists' conclusions about the nature of substances represent discoveries. In this section I return to the kind water.

Kripke (1980, p. 128) speaks for many discovery theorists when he un-hesitatingly asserts, "It certainly represents a discovery that water is $H_2O$." I think that, far from being certain, it is wrong to say that this was a discovery, even if we take for granted that 'water' refers to a microstructural kind. I think that we could have concluded after the arrival of modern chemistry that 'water' and '$H_2O$' have different extensions. Not only could we have so concluded, but this conclusion would not have accorded any less well with speakers' prior use of 'water' than did the conclusion that water is $H_2O$.

According to the familiar picture inherited from Kripke and Putnam, we learned that the extension of 'water' is the compound $H_2O$ by investigating samples of what we referred to as 'water'. We looked for the *distinguishing microstructural characteristics* of the majority of the liquid we called "water." Liquid bearing these characteristics of the majority of what we called "water" was really found to be water. The rest of what we may have called "water" was revealed to be some different kind of matter that we formerly mistook for water because of its apparent similarity.

Here is the reason I think that we could have concluded that water is *not* identical to the $H_2O$ compound: *The majority of what we prescientifically called "water" has more than one microstructural feature that we could have concluded distinguishes the true from the spurious samples believed to be water.* A feature other than being $H_2O$, perhaps one that overlaps with being $H_2O$, might have been taken to characterize what was to be called "water," and it would have been no more right or wrong to draw the conclusion that some other feature characterizes what is called "water" than to draw the conclusion that being $H_2O$ does the job.

To illustrate this claim, I offer the following story, in which two groups of English speakers conduct exploration into whether a certain sample of liquid is really water, really bears the relation *same microstructural kind* to what English speakers call "water." They come to different conclusions. My example involves a place similar to Putnam's famous Twin Earth, though there are important differences.

The time on Earth is just after the turn of the twentieth century. Scientists have learned a good deal about the physical universe by this time. Many elements have been discovered, as have been many compounds. It is held within the scientific community that the most basic matter in the world is the elements. Elements are distinguished by each one's having a different atomic number. Gold, for example, is the element of atomic number 79. It is also held that any two atoms of the same atomic number are type identical. So two hydrogen atoms will be indistinguishable, as will two oxygen atoms, and so on. And it is held that water is $H_2O$. All these conditions did obtain in fact.

Now I will part from history. A message reaches Earth from outer space. Intelligent inhabitants of a distant and undiscovered planet called "Deuterium Earth" invite a party of Earthlings to visit Deuterium Earth. A rare opportunity for one-way travel through space-time has chanced upon the cosmos. An opportunity to return will arise in thirty years or so.

So leaders on Earth decide to send a team of scientists on the journey. The voyagers prepare samples of many known kinds from the natural world to take to Deuterium Earth in order to compare them with material to be found there. A veritable Noah's ark is prepared with land and water animals and plants, as well as many chemical samples.

The journey takes place in 1905. I will focus on just one kind from Deuterium Earth that the scientists explore after their arrival: the kind of stuff on Deuterium Earth that resembles water.

The waterlike liquid on Deuterium Earth comes down in the form of rain, as water does on Earth. It fills oceans, lakes, and rivers and is drunk by Deuterium Earthlings. Soon after their arrival, Earthlings begin examining the waterlike liquid to see if it is really water – that is, to see if it bears the relation *same microstructural kind* to stuff Earthlings call "water." The two liquids look and behave very much alike when compared. But differences are soon noticed, even before molecular testing can be started. The first clue to the liquids' different structures comes when the goldfish tank from Earth is cleaned and the water is replaced with the Deuterium Earth liquid. The goldfish subsequently die. Other aquatic animals are tested for their reaction to the Deuterium Earth liquid. Both saltwater and freshwater animals find survival in the liquid impossible, though the liquid is not polluted. Crabs, shrimp, clams, and many fish types are tested and die. Aquatic plants die in the liquid as well. Not a single case of acclimatization is found. It is decided not to try to drink the liquid.

The next difference found is that Earth water and the Deuterium Earth liquid boil and melt at notably different temperatures. One cold day, ice sculptures of world leaders from Earth and from Deuterium Earth are carved out of ice from Earth water and Deuterium Earth liquid, respectively. A slight warming trend brings temperatures a little over 32 degrees F. The sculptures made of water brought from Earth begin to bead with water, while those made from Deuterium Earth liquid hold firm. The temperature rises to 36 degrees and above. The water sculptures melt entirely over time, but the Deuterium Earth statues do not even begin disintegrating. Even when the temperature is almost 39 degrees, the sculptures made from Deuterium Earth liquid resist forming droplets. It begins to snow. Despite the warm temperature, Deuterium Earth snow sticks and piles high, showing no sign of melting. At last the temperature rises again. It surpasses 40 degrees F and the frozen Deuterium Earth liquid begins to melt slowly.

Upon more in-depth investigation it is found that when the Deuterium Earth liquid is mixed with water samples from Earth, the two liquids can be separated with sophisticated scientific treatment. The scientists are puzzled; surely this stuff is microstructurally of a different kind from most of what we call "water." The mysterious structure making up the Deuterium Earth liquid is dubbed "PQR." Although it is at first believed that PQR is not found at all in the water samples brought from Earth, this turns out not to be so. Very small traces of PQR are later found in the scientists' water samples; ocean-water samples, for example, are found to contain about 0.015% PQR. (This is not tremendously surprising, because Earth water contains many foreign substances. Ocean water also contains small traces of gold, for example.)

Eventually, through sophisticated testing, it is learned that the Deuterium Earth liquid is, like the bulk of water samples brought from Earth, composed of an oxygen atom combined with two atoms that have, like Earth hydrogen, atomic number 1. But the Deuterium Earth atoms of atomic number 1 are very different from most atoms called "hydrogen" on Earth.

On Earth the great bulk of what is called "hydrogen" has a nucleus with one proton, or positively charged particle, making it atomic number 1. Here on Deuterium Earth, the same is true. But the stuff from Deuterium Earth has a neutrally charged particle (neutron) in the nucleus as well as a proton. So it has extra mass. At the atomic level it is structurally different from most Earth hydrogen. Contrary to previous beliefs, not all atoms of any given element, or atomic number, are type identical.

The scientists call the element without a neutron "protium." They call the element that has a neutron "deuterium," because it is found on Deuterium

Earth. Protium is found to be quite unlike deuterium in various respects. Some differences are small. Deuterium reacts chemically more slowly than protium. For example, deuterium reacts with chlorine at 32 degrees F more than thirteen times slower than protium does.

Big differences between protium and deuterium become apparent as soon as more serious communication can be achieved with the Deuterium Earth-lings. Earthlings are informed of an explosive weapon created with deuterium extracted from the Deuterium Earth liquid it largely composes. This weapon was used once at sea. The explosion caused a whole island to vanish. A hole was created 175 feet deep and one mile in diameter where the island had stood. No such bomb had ever been known on Earth. This deuterium weapon could not be made with Earth water's protium. So there are big differences between the two kinds of atoms that, with oxygen, make up the respective liquids.

In light of this information, the question is posed to the scientists: *Is $D_2O$ water?* Does it bear the *same microstructural-kind* relation to the majority of what we call "water"? Suppose in ancient Greece Archimedes had had a glass of $D_2O$ and called it "water" (in Greek). Would he have been right?

In the light of molecular differences between what is normally called "water" on Earth and $D_2O$, the scientists conclude that $D_2O$ is not true water. It does not bear the relation *same microstructural kind* to the majority of what Earthlings call "water." The scientists dub it 'dwater'.

Are the scientists right to say that $D_2O$ is not water? Putnam has something to say about a similar case. He supposes at one point in "The Meaning of 'Meaning'" that Archimedes has a piece of metal $X$ that he takes for gold, but that is not gold, because it does not bear the proper microstructural relation to the bulk of what we would call "gold." How could the absence of this relation be determined? Here is what Putnam has to say:

> Perhaps $X$ would have separated into two different metals when melted, or would have had different conductivity properties, or would have vaporized at a different temperature, or whatever. If we had performed the experiments with Archimedes watching, he might not have known the theory, but he would have been able to check the empirical regularity that '$X$ behaves differently from the rest of the stuff I classify as [gold] in several respects'.... This may not *prove* that it isn't gold, but it puts the hypothesis that it may not be gold in the running, even in the absence of theory. If, now, we had gone on to inform Archimedes that gold had such and such a molecular structure (except for $X$), and that $X$ behaved differently because it had a different molecular structure, is there any doubt that he would have agreed with us that $X$ isn't gold? (Putnam 1975e, pp. 237–8)

It seems that if we use the model Putnam has informally set before us, we should agree with our scientists that $D_2O$ is not water. It diverges in behavior from the bulk of what we call "water" on a number of accounts: $D_2O$ cannot support aquatic life, while water does. The two liquids have different melting and boiling points. They can be separated with proper scientific treatment. $D_2O$ can be used to build an extremely powerful bomb that cannot be made from ordinary Earth water. And the reason for $D_2O$'s deviant behavior on these and other points is its distinct molecular structure.

To return to the story, the thirty years on Deuterium Earth pass. It is time for the scientists to return to Earth. They take a tank of $D_2O$ back to Earth with them when they depart. Back on Earth in 1935, the returning pioneers cede the tank of $D_2O$ to the resident scientists, explaining that the tank does not contain true water as it appears to do but rather that it contains a newly discovered liquid, dwater, which serves on Deuterium Earth the role water does on Earth. The resident scientists, upon examining the tank, respond disappointedly that this is not a new liquid at all. The explorers have merely discovered an uncommon variety of water. It is composed not of ordinary hydrogen but of a hydrogen isotope called, coincidentally, "deuterium."

This concludes the story. All of the information about natural kinds such as deuterium and deuterium oxide is factual. Deuterium was discovered in 1931 on Earth. The weapon composed of it is the hydrogen bomb. And $D_2O$ is in fact considered water. Often $D_2O$ is called "heavy water" and normal water "light water."[8]

What do I think this story shows? I think it shows that we did not *discover* that deuterium oxide is water. Hence we did not discover that water is identical to $H_2O$. We could have concluded that some $H_2O$ (the variety that is $D_2O$) is not what we had been calling "water," as our space travelers concluded. I think the decision that just $H_2O$ made with protium bears the relation *same microstructural kind* to the majority of what we called "water" would have been no less acceptable a conclusion than that $H_2O$ bears the key relation. We cannot say that our space travelers were just flat wrong in concluding that $D_2O$ is not what they had been calling "water" and that we are just plain right in concluding that it is.

It might be suggested, on the contrary, that one community is right and the other is wrong, and that it is up to "final science" to determine which community has drawn the correct conclusion about the reference of 'water'. Final science has, at any rate, sometimes been urged to offer a principled settlement on disagreements over reference (see, e.g., Boër 1985, pp. 106, 114–16). Unfortunately, it cannot help here.

The reason final science cannot help is that the two communities agree on the physical facts about the world. Had the communities remained isolated, each could have independently reached this final science without either's having overturned its decision about the proper reference of 'water'. Indeed, it would be obvious to one from our community in the Final Science Age that the other community had arrived at the same scientific truths we had, if the other community were to present ours with a book of their ultimate science, along with the translational note (in Earthian English) that when 'water' appears in the book, the authors are not really talking about *water* but about protium oxide.

A more mundane example may more plainly show the futility of any appeal to final science here. Suppose I say to you, "Let me alter my idiolect a little: From now on, I'll use the word 'water' only for water made with protium, instead of for all water. I'll have a new term for all water: 'aqua'. You continue to call all water 'water'." Here it seems obvious that there is no substantial difference between us to be resolved by final science. But the differences between the two communities above seem relevantly similar (see also §V of Chapter 5). These communities have merely attached the term 'water' to different kinds.

The Deuterium Earth story shows how more than one microstructural kind may compete to be mapped to a natural-kind term from the vernacular. In the story, isotope kinds and element kinds compete. It should be noted that the moral of my story does not depend on both competing kinds' being equally fundamental[9]: The *red wolf*, *wolf*, and *canine* kinds vary in significance and priority, but all three are salient natural kinds. So if a scientist unfamiliar with canines were to point to a pack of red wolves interspersed by one or two wolves of another species and call attention to "the kind instantiated by those things, or the majority of them if there is not just one kind there," she would not have established determinate reference to any of the foregoing kinds. In the Deuterium Earth story, $H_2O$ and $D_2O$ were seen to present a similar situation. Ostension to matter containing much of one and a little of the other failed to single out either the $H_2O$ kind or the protium oxide kind. Given that both of these related, authentic, salient natural kinds were instantiated by the samples, no kind was seen to have an overpowering claim to have been the referent of 'water'.

$H_2O$ could not have been discovered to be the essence of what we have all along called "water" because there was another candidate, an overlapping but distinct kind that might as well have been said to be identical to what we called "water," by virtue of being relevantly like the majority of the samples of matter representing the kind. The natural world is full of such closely related kinds. To name just a few more cases: Sometimes a substance of one chemical composition will be molded into different shapes and forms at the microstructural level. This may affect surface properties such as color or hardness, as in the case of opal, some of which is arranged at the submicroscopic level into uniform, ordered spheres packed together and some of which is not. Suppose the former is baptized, say, "opal." Should the latter belong to this baptized kind? We could say yes. 'Opal' has been determined to embrace both. But we could say no. We *have* said no, in like cases.[10] Overlapping kinds can be found at more fundamental levels. In addition to the obvious same-kind relations between substances composed of the same elements, there are also some between substances composed of different elements. Tourmaline ($XY_3Z_6B_3Si_6O_{27}(OH)_4$), for example, contains various combinations of elements in the first three places of its formula. Each combination is a distinct kind. But the varieties form a collective kind as well. Different isotopes present us with just one example of microstructures that are related but distinct and that might or might not be lumped into a kind anchored by samples. The world invites a lot of decision in determining whether this or that bit of matter bears the *same structural kind* relation to a sample group.

Perhaps there are many historical cases for which the *extension* of a vernacular term has been clear since pre-scientific times. Not all cases are likely to have been like 'water' in that more than one possible extension might have been matched to a kind term. But even if a kind term's extension is clear, and even if that extension differs from other actual substances by having some salient, scientifically recognized microstructure, the relevant kind is not likely to be *identical* or straightforwardly identical to that microstructure. Even when the actual world cooperates by carving substances into distinctive microstructural camps, other possible worlds are likely to present microstructures that are related to but distinct from any that we might specify, so that we will again find vague cases.

Consider a counterfactual world in which, to adapt an example from the literature (Steward 1990, pp. 389ff.), people discover $H_2O$ with a special sort of proton in the hydrogen, causing the substance to be pink and fluffy. Would we call this "water"? The author of this example thinks we would not, and I agree that we might not. I think we could go either way. I think that it will never be determined that the substance does or does not straightforwardly belong to the extension of 'water', though it is $H_2O$, unless we find some and the need arises to decide the matter.

Take a situation in which we discover a close relative of $H_2O$. Suppose a few of our northern lakes are found to contain something almost indistinguishable from $H_2O$, but that does not contain true hydrogen, because the protons in the hydrogen atoms have been replaced by a surrogate particle. We might call such a substance "water," though it is not $H_2O$.

We do not know that such fanciful situations are metaphysically possible. But nor do we know they are not. As far as we know, it is metaphysically possible that we could find a substance revealing more open texture, so we do not know that 'water' and '$H_2O$' refer straightforwardly to the same items in all possible worlds.[11] And speakers' conclusion that the new substance does or does not merit the older name 'water' would refine the language: It would not be a discovery.

<br>

### IV. LOOKING AHEAD

I have argued that scientists have not discovered that 'Water $= H_2O$' is true and its denial false. But, some might wonder, would it not be *possible* to have a posteriori discoveries of this kind? It seems reasonable to think that such discoveries would be possible, as I have indicated in an earlier chapter (Chapter 1, §I.4). Let us assume that scientists' empirical information about the world is right. Assume, in particular, that there are no underlying confusions in chemists' notion of a chemical composition. Assuming that there are none, scientists today might succeed in coining a term in more or less the way Kripke and Putnam think ordinary speakers have always coined terms. Scientists might point to a glass of pure $H_2O$, before they know that it contains $H_2O$, instead of something else like rubbing alcohol, and say "We use 'aqua' for the chemical element or compound exemplified in that glass." If such a dubbing ceremony were to occur, then, after a little empirical investigation, it would seem right to conclude that the sentence 'Aqua $= H_2O$' is a true statement about aqua's essence. There would be no worry about anything like XYZ counting as aqua, despite XYZ's superficial similarity. XYZ would not count as aqua because the dubbers have specified that the term 'aqua' refers *just to a particular compound or element*, and XYZ is a totally different compound. $D_2O$ would not create trouble, either: It would simply count as aqua. Chemists have consciously specified that the term 'aqua' is to refer to the chemical *compound or element* instantiated by a certain liquid, and *not* to an isotope-limited kind.

The possibility of a posteriori discoveries of essences seems *coherent*, then, even if natural language does not cooperate with the Kripke-Putnam picture. The orthodoxy that 'Water $= H_2O$' was discovered to be true by empirical investigation seems wrong. But the limited claim that there *could be* statements about kinds' essences that are discovered to be true by empirical investigation seems right.

One might find reason in the present chapter and the preceding chapter for doubting even that there *could* be discoveries of the foregoing sort, and because of the importance of these doubts, I will return to discuss them in Chapter 6. But in the following chapter, I set aside the question of whether or not there could be discoveries of essence. Whether scientists *could* announce the discovered essence of a kind or whether they could not, scientists' conclusions about essences are very often *not* discoveries. Scientists' conclusions represent some theory change but also some meaning change. Conceptual change that blends meaning change and theory change like this lies at the center of famously knotty philosophical issues, including the issue of incommensurability. That is the topic of the next chapter.

In Chapter 6, I continue to discuss conceptual change that blends meaning change and theory change. There I will have more to say that is pertinent to the question of whether there could be discoveries of essence. Although a few paragraphs ago I argued that there could be such discoveries with respect to chemical kinds, and I would be prepared to defend the same claim about biological kinds, one might, as I have said, be encouraged by this chapter and the preceding chapter to conclude otherwise. These chapters might encourage doubt as to whether there even could be statements about kinds' essences that are discovered to be true by empirical investigation, because they might encourage doubt about whether there is any difference between a meaning change and a theory change. Without such a difference, there could be no pure discovery of a kind's essence, because if there were ever to be such an essence discovery, it would have to be a mere change in theory about the kind; it could not be a change in the relevant kind-term's meaning. In Chapter 6, I address the issue of whether there is a difference between meaning change and theory change.

# 5

# Linguistic Change and Incommensurability

In the past two chapters I have argued at some length that our use of natural-kind terms has changed over time as our scientific sophistication has increased. We continually refine our use of these terms in response to scientific research. For example, earlier speakers used the word 'fish' differently from the way we do. So we have not simply discovered that earlier speakers erred in accepting the sentence 'Whales are fish'. Rather, we have changed what 'Whales are fish' means.

This sort of instability in language use arising from scientific inquiry may seem to cast doubt on whether there is rational progress in science. It may seem to suggest that today's scientific community is not really better informed or more advanced than earlier speakers in the investigation of kinds addressed by earlier speakers, for speakers today really address kinds *other* than those addressed by earlier speakers. Scientists of different periods *appear* to be studying the same kinds because later scientists retain the older kind terms of earlier scientists; but appearances mislead, because as time progresses, the older words lose their original meanings and come to mean something else entirely. This skeptical worry calls to mind Thomas Kuhn, Paul Feyerabend, and other proponents of the so-called "incommensurability thesis," who insist that conceptual change is marked by linguistic change.

The causal theory of reference to natural kinds, as championed by Putnam and Kripke, is supposed to provide an *answer* to the threat of language instability posed by incommensurability theorists. The causal theory of reference is supposed to assure the *continuity* of meaning and reference over time. This continuity is supposed to ensure that true sentences about natural kinds that were discussed in earlier stages of scientific development have been discovered to be true, not stipulated to be true. There is no change of meaning, so no incommensurability.

I have argued in the past two chapters that our natural-kind terms do change in meaning, as science progresses. The causal theory of reference seems roughly right, but it does *not* secure linguistic stability.[1] Because it does not secure linguistic stability, the causal theory cannot secure the objectivity of science in the way supposed. I discuss this point in detail in section (I) below. The apparent threat of incommensurability requires a different response. I develop such a response in the remaining sections.

### I. THE DESCRIPTION THEORY AND THE CAUSAL THEORY: NO PROBLEM, NO SOLUTION

There are different types of incommensurability. The earlier literature on the topic, in particular, sometimes discusses "methodological" and "observational" incommensurability as well as "conceptual" or "linguistic" incommensurability. What is at issue in this chapter is conceptual or linguistic incommensurability, which is supposed to arise as a result of changes in the use of key terms as science advances. Conceptual or linguistic incommensurability, hereafter just "linguistic incommensurability," is the most widely discussed variety of incommensurability. Part of the reason for the greater attention paid to this variety of incommensurability is that it is this variety that Kuhn later comes to identify as what he was really after all along in his discussion about incommensurability:

> My original discussion described nonlinguistic as well as linguistic forms of incommensurability. That I now take to have been an overextension resulting from my failure to recognize how large a part of the apparently nonlinguistic component was acquired with language during the learning process. (Kuhn 1990, p. 315n.)[2]

In addition to being the variety of incommensurability that the mature Kuhn considers important, linguistic incommensurability is also the variety of incommensurability that the causal theory of reference is supposed to block. Its connection to the causal theory of reference helps to explain why more attention has been paid to linguistic incommensurability than to other varieties of incommensurability. Its connection to the causal theory of reference also explains why linguistic incommensurability is the variety of incommensurability that is of interest here.

In the present section I employ one of Kuhn's examples to illustrate that linguistic instability and all of the worries attending it arise whether we accept the causal theory of reference or the description theory of reference, which the causal theory has replaced.[3] I begin with a closer examination of the worries attending instability.

## I.1.  *The Apparent Threat*

As I have said, incommensurability theorists insist that major scientific transition brings important changes in the use of terms. Because terms undergo this change in use, Kuhn insists, "Communication across the revolutionary divide is inevitably partial" (1970a, p. 149). He offers the term 'mammal' as an example of a term from biological theory whose meaning has shifted. This term, he suggests, changed meaning when systematists discovered and attempted to classify the monotremes, whose living representatives include just the platypus and the echidna. Monotremes have both reptilian and mammalian features. When systematists learned about the monotremes, they classified the new organisms as mammals. Still, Kuhn says, "to be a mammal is not what it was before" (1990, p. 307).

Like Kuhn, Feyerabend also recognizes linguistic and conceptual change:

> [I]ntroducing a new theory involves changes of outlook both with respect to the observable and with respect to the unobservable features of the world, and corresponding changes in the meanings of even the most 'fundamental' terms of the language employed. (Feyerabend 1981a, p. 45)[4]

Is the linguistic and conceptual change described by Kuhn and Feyerabend *itself* incommensurability, or does it *cause* incommensurability? That depends on what (linguistic) incommensurability is, and about that there is disagreement: Hence, no simple answer is available. I address the question later (in §IV below). There is no need yet to address it or its attending complications, which do not matter here: What is important for now is that the relevant instability, which is or which causes incommensurability, appears to undermine scientific progress for familiar reasons. Such instability suggests that the appearance of progress is an illusion fostered by a series of puns. When scientists appear to augment and replace earlier claims about a given kind, they have in fact changed the subject. There is not the kind of rational accumulation of knowledge about a continuous object of study that there appears at first glance to be. Indeed, the worry arises that there are no rational grounds for theory change: As earlier theorists have not been corrected by later speakers, later theories are not closer to the truth or rationally superior.

Certainly Kuhn and Feyerabend have fanned the fears expressed above. After quoting a passage from the *Origin of Species* in which Darwin confesses, "I by no means expect to convince experienced naturalists [but rather] look with confidence to the future – to young and rising naturalists," Kuhn famously remarks that such passages "have most often been taken to indicate that scientists, being only human, cannot always admit their errors, even when confronted with strict proof. I would argue, rather, that in these matters neither proof nor error is at issue" (1970a, p. 151). Kuhn's suggestion that Darwin's opponents cannot be accused of error is hard to reconcile with the idea that Darwin made progress. Feyerabend writes that incommensurable theories are not rationally comparable: "Their *contents* cannot be compared." The usual means of accounting for scientific rationality are inapplicable. "What remains are aesthetic judgements, judgements of taste, metaphysical prejudices, religious desires, in short, *what remains are our subjective wishes*" Feyerabend says (1988, p. 226). Not surprisingly, many philosophers are unhappy with these conclusions.[5]

## I.2. *Description Theories Lead to Incommensurability*

With the rational progress of science apparently on the line, some measure to block instability has seemed urgent. Many have diagnosed the description theory of reference as the source of trouble and have prescribed its replacement by the causal theory of reference as the remedy. The description theory has seemed to be the source of trouble because it does readily lead to referential change, because according to the description theory, the descriptions associated with a term are what determine its reference, and they change over time. When descriptions associated with a term are dropped, augmented, or altered, a term changes in meaning, and perhaps reference as well. Many have pointed out that the description theory of reference leads to incommensurability in this way.[6]

There are different versions of the description theory. Some lead to more radical changes in meaning and reference than others. Feyerabend seems to endorse a version of the description theory that yields radical changes of meaning and reference.[7] When there is a change in language use, Feyerabend says, it is extreme:

[W]e shall diagnose a change of meaning either if a new theory entails that all concepts of the preceding theory have zero extension or if it introduces rules which cannot be interpreted as attributing specific properties to objects within already existing classes, but which change the system of classes itself. (1981b, p. 98)

In the event of meaning change, older terms do not refer at all in the language of the new theory, or else they are no longer used with the same reference in the language of the new theory.

Evidently Feyerabend does not take the earlier use of terms to be unclear or vague, as I have argued it was in Chapters 3 and 4. Rather, on his view earlier use was perfectly clear: Such terms straightforwardly referred to different kinds than they refer to today. This version of referential change is too severe to be plausible, because, as various philosophers have observed, it does not allow for any continuity at all through conceptual change (see, e.g., Kitcher 1978, p. 546n.; Sankey 1994, p. 160).

Feyerabend's version of the description theory yields changes of reference that are too drastic. But not all description theorists are committed to Feyerabend's implausible conclusions about drastic reference change. Proponents of the *cluster-of-descriptions* theory of reference, in particular, are not committed to any such drastic shifts of reference. Kuhn suggests such a theory in various places. Consider his example 'mammal'. On the cluster theory, speakers from centuries past would have referred by 'mammal' to whatever satisfies enough of the descriptions speakers associated with mammals, or enough of the most important descriptions. These would have included the descriptions 'live-bearing', 'lactating', 'hairy', and so on. Whatever possessed all of the descriptions clearly belonged to the extension, and whatever possessed none of the descriptions clearly failed to belong to the extension. But as Kuhn indicates, a group of organisms could meet some of the descriptions and not others: This allows for the discovery of borderline cases, too, which reveal vagueness in the term 'mammal'. "What should one have said when confronted by an egg-laying creature that suckles its young? Is it a mammal or not?" There was no simple yes or no answer, Kuhn says. It is a question that the users of the term 'mammal' did not anticipate having to answer. Cases like this spur theorists to refine the use of their terms to reduce vagueness. "Such circumstances, if they endure for long, call forth a locally different lexicon, one that permits an answer, but to a slightly altered question: 'yes, the creature is a mammal' (but to be a mammal is not what it was before)" (Kuhn 1990, pp. 306–7).

The cluster-of-descriptions theory of reference recognizes a change of meaning in this case, because the descriptions associated with 'mammal' have changed. The description 'live-bearing' was associated with the term 'mammal' before the discovery of the monotremes, but not after. But this is a gentler change of meaning than the one Feyerabend recognizes. Because the monotremes satisfy some but not all of the descriptions formerly associated with 'mammal', they were neither clearly in nor clearly out of the extension before speakers made a decision to call them "mammals." There is, therefore, some needed conceptual continuity, as well as change.

## I.3. *The Causal Theory's Failed Solution*

It is easy to see how examples with some but not all of the properties associated with 'mammal' could present taxonomic problems like the foregoing, given the descriptive theory of reference. So it is easy to see how proponents of the descriptive theory would be driven to recognize meaning shifts in theoretical terms. Fortunately, it may seem, proponents of the *causal* theory of reference are not committed to such linguistic instability. On the causal theory, reference is determined not by theory-laden descriptions that speakers associate with a word but rather by the way the *world* is. Speakers' words are causally linked to the world by way of paradigmatic samples. 'Mammal' refers to whatever has the essence of dogs, horses, and other samples used to coin the term in a baptismal ceremony. Use of the word depends on the nature of those samples in the world.

So according to the causal theory, the descriptions speakers associate with the term 'mammal', like 'live-bearing rather than egg-laying', could change without there being any change in meaning or reference: Only the nature of the original paradigmatic samples used to coin 'mammal' determines meaning and reference. Accordingly, it is up to scientists to investigate those samples in order to find out what their nature is, and hence to what things the term 'mammal' refers. Scientists do not change the meanings of terms like 'mammal' when they report findings about the essences of the relevant kinds; they just inform other speakers about what does or does not belong in the extensions of the terms as speakers have been using them all along. Earlier speakers and present-day speakers refer to the same kinds by the use of their terms. There is no linguistic change, and incommensurability is blocked.

Putnam employs the causal theory to undermine incommensurability in this way in several highly influential discussions. The causal theory allows Putnam to insist that today's scientists offer "*better* descriptions of the *same* entities that earlier theories referred to" (1975e, p. 237). Putnam's response has received wide acceptance. Shapere observes that the causal theory of reference has provided, in recent years, "the most influential approach to the incommensurability claim" (1998, p. 735). Neale (2001, p. 13) remarks on the "profound impact" that this response to incommensurability has had.

Papineau (1996, pp. 3–4) reports that the introduction of the causal theory has effectively silenced a once-lively debate over the issue of meaning invariance. Many other writers have offered similar testimony to the great power and influence of Putnam's use of the causal theory against incommensurability.[8] The causal theory is widely credited with resolving a vexing problem.

Unfortunately, the causal theory does not live up to its promise. The theory is, contrary to wide acclaim, useless in blocking instability. The causal theory of reference leaves room for plenty of reference change, as the past two chapters indicate. The reason that the causal theory allows for reference change is that causal baptisms, which according to the causal theory endow terms with their reference conditions, are performed by speakers whose conceptual development is not yet sophisticated enough to allow the speakers to coin a term in such a way as to preclude the possibility of open texture, or vague application not yet recognized. When speakers recognize that their use of a term is vague, they tend to offer further specification for its use. That further specification amounts to a stipulation that changes the term's meaning.

The causal theory does not, then, significantly differ from the cluster-of-descriptions theory of reference with respect to linguistic change. On both theories the use of a term is refined after empirical investigation reveals unwelcome open texture in the earlier use of the term. Recall that on the cluster account, 'mammal' has had its use altered somewhat with the discovery of the platypus and the echidna, borderline animals that did not clearly belong or clearly fail to belong to the group that earlier speakers called "mammals." The same holds on the causal account.

On the causal account, speakers at a museum in Europe some centuries ago might have coined 'mammal' by pointing to a horse, a dog, a bear, a human, and so on, and declaring "Mammals are things with the unknown essence of these." Of course, they might have pointed to a few foils as well, like snakes, fish, and birds: "These are *not* mammals." Speakers would have secured the "mammalian" status of dogs, horses, and other mammals known to speakers at the time of coining had they performed such a baptism. One might expect that speakers would also have secured the "mammalian" status of the undiscovered platypus: Because it would have shared the essence of those animals serving to ground 'mammal', no meaning change would have occurred when the platypus was later pronounced a mammal. Modern scientists, who are supposed to tell us whether the platypus has the right essence, recognize phylogenetic, or historical, essences. Perhaps scientists have found that the platypus is a "mammal" in the original sense of 'mammal' because the platypus has the right evolutionary history.

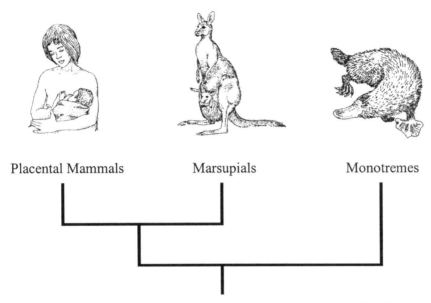

Placental Mammals      Marsupials      Monotremes

**Figure 5.1.** A cladogram of the mammals. Adapted in part from M. Strickberger, *Evolution*, 2nd edition. Copyright 1996 by Jones and Bartlett Publishers, Sudbury, Massachusetts. www.jbpub.com. Reprinted with permission.

This happy picture of discovery does not withstand scrutiny. There are, as I have observed (in Chapter 3), intermediate locales on the genealogical tree. The platypus, along with the other monotremes, occupies just such a locale. There is a branch including all mammals except the monotremes, and this is embedded in another, larger branch that extends just as far as the monotremes (see Figure 5.1). Both branches are legitimate genealogical units, but it is not clear which is to be identified with the "mammals" of earlier speakers. Because one but not the other of these branches includes the monotremes, it is not clear whether the monotremes are "mammals." Speakers are left to decide whether to call the monotremes "mammals" on the basis of whether they have the appropriate features, which, as I have indicated, seem to leave the matter open. Speakers have chosen to call the monotremes "mammals." But speakers could have said that they are not "mammals," concluding that they branched off before the lineage leading to the "mammals." Either decision alters the original use of 'mammal'. That term's use had not been determined one way or the other by earlier speakers.

Kripke and Putnam imagine that what unites the mammals or any other such group is some sort of genetic essence, rather than an evolutionary essence. Practicing systematists do not classify groups like the mammals by genetic essences, although they *do use* genetic information to *determine* the evolutionary relationships by which organisms are classified. In any case, the suggestion that essences are genetic would not secure meaning stability. Genes are able to help determine evolutionary relationships only because they change gradually, just as traditionally observed characteristics do: Closeness in phylogenetic relationships is thus typically reflected in the degree of genetic materials' similarity. For that reason, phylogenetically intermediate groups ought to be genetically intermediate as well.[9] So whether underlying essences are genetic or phylogenetic, it should be equally unclear whether the monotremes have the "mammalian" essence.

The use of 'mammal' seems to have changed with the discovery of the monotremes. The causal theory of reference does not change that. It does not preserve the linguistic stability of 'mammal' any more than the cluster-of-descriptions theory does. Nor does the causal theory preserve the stability of 'fish', 'dinosaur', 'reptile', or a host of other terms. For various reasons, each of these terms was marked by a vagueness that was later exposed by scientific investigation. After the vagueness was revealed, speakers refined it out of the words, altering their meaning.

In sum, then, the causal theory of reference does not assure linguistic stability. Nor does its hoary competitor, the description theory, or, more precisely, a suitably moderate cluster version of it, even threaten drastic reference shifts evaded by the causal theory. Both theories suggest gentle shifts in terms' use. The causal theory simply does not deliver on its promise to rescue meaning stability from the wreckage threatened by earlier work on reference. The common wisdom that the description account is the problem behind incommensurability and the causal theory the solution is erroneous.

## II.  CHANGE THROUGH DARWIN'S REVOLUTION

The causal theory's achievements have been substantially overestimated. This theory fails to rid us of the incommensurability problem. That important problem remains to be addressed.[10] Now that I have argued that the causal theory of reference cannot block instability, and that instability arises with or without the causal theory, it is time to set aside the causal theory and focus just on instability. In the following examples, instability arises because reference is established with the help of descriptive information that changes over time. How these examples might suggest limits on the causal theory's range of application or a need to modify that theory will not enter the discussion explicitly.

The chief question for the rest of this chapter is whether language instability, which seems to be here to stay whatever our theory of reference, undermines the progress of science. Most of us have a strong inclination to say that science does make progress, but the question is whether and how this commonsense view is tenable given linguistic change. Here I will focus on scientific revolutions, which lie at the center of Kuhn's work. More specifically, and in keeping with the general focus on biological kinds in earlier chapters, I will focus on scientific revolutions in biology. I will defend a number of moderate Kuhnian theses concerning linguistic instability. Still, I will argue that when the linguistic change is properly characterized, it gives us no reason to doubt that there is scientific progress through revolutions.[11]

### II.1. *Languages and Communication Through Darwin's Revolution*

The Darwinian revolution replaced special creationism with evolutionism. In the Introduction to the *Origin*, Darwin introduces his conclusion

> that the view which most naturalists entertain, and which I formerly entertained – namely, that each species has been independently created – is erroneous. I am fully convinced that species are not immutable; but that those belonging to what are called the same genera are lineal descendants of some other and generally extinct species. . . . (1859, p. 6)

In expressing his thesis, I will argue, Darwin changes the meaning of the crucial term 'species'. If my arguments are successful, then Kuhn is right in supposing that the use of 'species' has undergone a major transition with the development of science (Kuhn 2000a, p. 85). Similar changes seem to attend other taxonomic terms, like 'variety', but I will focus on the central term 'species'.

Before Darwin effected a change in their general use, scientific terms like 'species' were laden with creationist theory. Speakers intended to use 'species' for groups that include a specially created original population of organisms, such as a first pair, and all of the original population's genealogical relatives. An idealized causal baptism performed to coin the term 'species' might have proceeded more or less as follows: "A species is any group like *that* and all of its blood relatives" (the speaker points to an organism). Speakers would have been comfortable conveying the use of 'species' this way as well as by enumerating examples: the radish species, the polar bear species, and so on. Darwin was certainly aware of the influence of creationist theory on his fellow naturalists' use of 'species'. Although he emphasized that the naturalists of his day disagreed over how to use 'species', he acknowledged that they tended to agree that the term was to be used for specially created groups. "Generally the term includes the unknown element of a distinct act of creation" (1859, p. 44).

Because the language of Darwin's contemporaries reflected creationist ideas, Darwin's position that new species have been generated by evolutionary change sounded paradoxical to many of his critics. The entomologist Thomas Vernon Wollaston, for example, charges that by virtue of the "*idea* involved by naturalists in the term 'species,'" what counts as a species is a group united by

> blood-relationship acquired by all the individuals composing it, through a direct line of descent from a common ancestor; and therefore, it is no sign of metaphysical clearness when our author [Darwin] refuses to acknowledge any kind of difference between 'genera,' 'species,' and 'varieties,' except one of *degree*. (1973, p. 129)

Wollaston seems to suppose that by muddling his concepts, Darwin fails to see that 'Species do not arise by evolution' is true *by definition*. Such a reaction was not uncommon, and it provides grist for Kuhn's mill. The reaction illustrates that Kuhn is right not only to recognize a shift in language but also to recognize that sentences rejected by one community may behave, for another community, "very much like a purely logical statement that no amount of observation could refute" (1970a, p. 78; see also pp. 149, 184; 1981, p. 9).

Not surprisingly, such variance in the use of key theoretical terms could strain communication on the subject of whether species arise by evolution. The distinguished botanist Hewett C. Watson acknowledges the trouble openly. "This will be a difficult subject to treat, because the very definition of the term 'species,' as usually given, involves an assumption of non-transition," Watson writes before the *Origin* (1845, p. 147). Darwin and his allies would soon become frustrated over being accused of ignorance about the meanings of central terms like 'species' and 'variety' (Hull 1973, p. 141). Darwin expresses his frustration frankly in a letter written in 1860: "I am often in despair in making the generality of *naturalists* even comprehend me. Intelligent men who are not naturalists and have not a bigoted idea of the term species, show more clearness of mind."[12] Darwin here testifies to the truth behind another Kuhnian maxim, that communication across the revolutionary divide is only partial.

Precisely because communication is only partial, Kuhn's initially startling further claims that "neither proof nor error" is at issue, and that past opponents of today's orthodoxy were not just wrong, turn out to enjoy more merit than a first glance suggests. By the very meaning of 'species' on pre-revolutionary speakers' use, it is doubtful whether any amount of evolution could generate a new "species." Pre-revolutionary speakers would have expressed a dubious claim by the sentence 'New species appear by evolution', so their rejection of the sentence was not entirely wrong, even if evolution has generated all of today's organisms from a common primitive ancestor.

Earlier speakers who deferred to the established use of 'species' were not quite wrong to reject 'New species arise by evolution'. But they were not quite right, either. Rather, 'species' turned out to be vague. Before the acceptance of evolution, speakers tacitly intended to use 'species' to refer just to the various groups of organisms that are relevantly like those that naturalists *called* 'species'. Given such a use, there would be many distinct "species." Polar bears would be members of one species, giraffes would be members of another species, and radishes would be members of still another species. Speakers also intended for 'species' to refer just to groups including all genealogical relatives of an original population of organisms. These intentions turned out to conflict, because the groups said by naturalists to be distinct "species" fail to include all genealogical relatives of any original population. The conflict reveals an unclarity about what, if anything, 'species' denotes in the discourse of pre-revolutionary speakers.

Because of the vagueness in 'species', it is not clear whether 'New species arise by evolution' expressed a conceptual falsehood or an unrecognized truth. Individual speakers refined the vagueness out of 'species' in order to yield a clear answer to whether evolution renders 'New species arise by evolution' true or false. For Darwin the sentence became true, because he opted to employ 'species' for groups that, like the polar bear or the radish, naturalists had *called* "species" on the assumption that they are specially created groups.[13] Darwin's newly refined use of 'species' *preserves* the intuition or intention of earlier speakers that a sufficient condition for the application of 'species' should be that a group be relevantly like groups that naturalists called "species." However, Darwin's proposed use *violates* the intention of earlier speakers to use 'species' in such a way that a *necessary* condition for its application is that a group include all genealogical relatives of an original population. The polar bear species does not include all genealogical relatives of its founding population: The species does not include radishes, which are related genealogically to polar bears.

Certainly Darwin's new use of 'species' is one sensible proposal for substituting a vague use for a more precise one after conceptual disruption, in a way that accords more or less with earlier use. But some of Darwin's contemporaries offer different proposals for how 'species' should be used after the disruption. These other proposals also accord more or less with earlier use.

Consider the proposal suggested by William Hopkins, a mathematician and critic of Darwin:

> Every natural species must by definition have had a separate and independent origin, so that all theories – like those of Lamarck and Mr. Darwin – which assert the derivation of all classes of animals from one origin, do, in fact, deny the existence of natural species at all. (1973, p. 241)

According to Hopkins, 'species' fails to refer if evolutionism is true. Hopkins's proposed use of 'species' honors certain intentions of earlier speakers. It honors the intention, first, to restrict the use of 'species' to groups that include an original population and all of its relatives. Second, Hopkins's use of 'species' honors speakers' intention to restrict the use of 'species' to groups that are roughly like the kinds of groups naturalists *call* "species," such as the polar bear, the giraffe, and the radish. Nothing meets both of these conditions: Groups like the polar bear, the giraffe, or the radish do not include an original population and all of its relatives. Therefore, 'species' fails to refer.

Although Hopkins's proposed use of 'species' honors some intentions of earlier speakers, it fails to honor others. It fails to honor the intention to count the polar bear, the giraffe, the radish, and so on as "species." Being relevantly like the polar bear or the giraffe was supposed to be a *sufficient* condition for being called a "species." But on Hopkins's proposed use, it is not: 'Species' does not refer to the polar bear or the giraffe or anything else. Like Darwin's proposed use, Hopkins's proposed use reflects some intentions of speakers but not others. It amounts to a refinement in the use of 'species'.

Hopkins's response to Darwin's use is anticipated by J. E. Bicheno, who proposes, several decades earlier in a talk addressed to the Linnean Society of London, that naturalists should respond in essentially this way in the event that evolutionism should be found true.

> We have agreed that a species shall be that distinct form originally so created, and producing by certain laws of generation others like itself. . . . There is this inconvenience attending the use of it by naturalists, that it assumes as a fact, that which in the present state of science is in many cases a fit subject of inquiry; namely, that species, according to our definition, do exist throughout nature. It is too convenient a term to be dispensed with, even as an assumption; only care should be taken that we do not accept the abstract term for the fact. (1826, pp. 481–2)

Like Hopkins, Bicheno is prepared to deny the existence of "species" if the creationist presuppositions determining the use of his term should turn out to be wrong.

English had both its Darwins on the one hand, and its Hopkinses and Bichenos on the other. Both parties seem to refine the language, though in different ways. Neither side's proposal seems to accord perfectly with earlier use. Given the truth of evolutionism, one might have affirmed 'New species evolve from more primitive ones' or one might have affirmed 'Species do not exist; rather, the different alleged species are related by evolution'.

Nor are the foregoing two proposals for refinement exhaustive. On a third proposal, speakers should affirm *neither* 'New species evolve from more primitive ones' *nor* 'Species do not exist'. Rather, speakers should say that there is now and always has been just *one* "species," or at most only a few. They should affirm, 'All animals belong to the same species, having descended from a single ancestor'. On this use, whales, spiders, and humans all belong to one "species," given evolution. Darwin's supporter Henry Fawcett complains shortly after the publication of the *Origin* that anti-Darwinians used 'species' in this way. On such a use,

> the descendants of any varieties of a particular species must always be con-sidered as belonging to this species, and . . . however much in succeeding ages such descendants may differ from the parent stock, this difference can never entitle them to be ranked as distinct species. . . . (Fawcett 1973, p. 279)

This use honors the intention of earlier speakers to use 'species' in such a way that all blood relatives of a lineage's original population belong to its "species" but discards the intention to use 'species' in such a way that there are lots of "species." Groups that naturalists have *called* "species" turn out not to be real "species," because they are too finely distinguished: There are lots of them. Darwin's proposed use is just the reverse: For Darwin the supposition that there are lots of "species" is retained, and the supposition that all blood relatives of a lineage's original population belong to that "species" is discarded. This other proposed use seems to accord as well as Darwin's with the pre-revolutionary use of 'species', as Watson testifies some years before the publication of the *Origin*:

> The very definition of the term "species," as usually given, involves an as-sumption of non-transition; so that any case of real transition – supposing such a case to be adduced – would be set down simply as evidence to disprove the duality of the species. (1845, p. 147)

Indeed, writes Watson, "the widest change ever seen, in the descendants of any plant or animal, would only entitle them to the name of 'variety,' according to recognized usage in the application of these terms" (1845, p. 111).

Hints that 'species' would, in the event of evolution, apply to all blood relatives of any member go much further back in time. The eighteenth-century evolutionist Charles Bonnet, for example, had allowed for the possibility of massive change in a lineage but would have balked at the thought of species' originating that way. According to Bonnet, as Bentley Glass relates, "Plants might rise to the state of animals, oysters and polyps to that of birds and quadrupeds, monkeys to that of men, and men to that of angels; but for each it would be a metamorphosis, not a transmutation of species" (Glass 1968, p. 168).[14]

## II.2. *Does Language Force Disagreement?*

The term 'species' was vague before the Darwinian revolution. Its use might have been refined in different ways. Darwin's opponents, who resisted evolutionism, often rejected Darwin's refinement, accepting instead a use of 'species' according to which 'New species arise by evolution' comes out false. Wollaston and Hopkins, for example, both resisted Darwin's evolutionism and his refined use of 'species'. On their proposed use of 'species', 'New species arise by evolution' could not possibly be true, however the various types of plants and animals came into being.

This phenomenon of substantial disagreement coinciding with different uses of terms might suggest that language forces disagreement. Kuhn suggests such a worry when he says that scientists on opposite sides of a revolutionary divide are able to offer only circular arguments in their own support.[15] Fawcett raises much the same worry about circular reasoning in a review of the *Origin* (Fawcett 1973, p. 290). Fawcett supposes that creationists' very use of the word 'species' commits them to miraculous acts of creation.

How could the use of a term commit anyone to a substantive thesis? Fawcett explains the matter thus: He begins by observing that it is no longer a matter of controversy that new species have been introduced over time. That new species have been introduced is clear to everyone, a "demonstrated truth" (p. 289). As Fawcett writes,

> If we go back to a comparatively modern geological epoch, it will be found that all the fossil animals belong to undoubtedly distinct species from any which exist at the present time. This is admitted by every naturalist and geologist, whatever may be his opinion on the origin of species. (p. 279)

Yet, Fawcett goes on to say, if 'species' is used as creationists use it, so that "the descendants of any varieties of a particular species must always be considered as belonging to this species," then one cannot offer an evolutionary account of the origins of the new species that have undeniably arisen. "It therefore becomes necessary to suppose that the same effort of Creative Will, which originally placed life upon this planet, is repeated at the introduction of every new species; and thus a new species has to be regarded as the offspring of a miraculous birth" (1973, p. 279).

Contrary to Fawcett's suggestion, however, a creationist's definition of 'species' cannot force her to accept miracles. A creationist's definition could be accepted with equanimity by evolutionists. It is just that an evolutionist who accepts a creationist's use of 'species', according to which all descendants of any species must by definition belong to it, should *deny* the "demonstrated truth" that "new species" have been introduced continually over time. The evolutionist should agree that new kinds of plants and animals have arisen over time but should not agree that these are new *"species,"* given the relevant use of 'species'. Given that use, evolutionists should say, "The many living forms around us, like humans and radishes, evolved from more primitive organisms, so they are not distinct species after all." In this way evolutionists deny that there have been miraculous acts of special creation without assenting to anything like an analytic falsehood. Creationists will reject evolutionists' newly phrased claim, of course, but they cannot claim that it fails *by definition*.

Hopkins is more clear headed about this issue than Fawcett. Hopkins seems to be aware that differences between his use of a word and Darwin's use of it do not force him to disagree with Darwin. It forces him only to express the issues differently than Darwin does. Hopkins expresses the primary issue raised by Darwin in his own terms. As Hopkins puts it, the issue is about whether "species" exist. Darwin's theory denies that "species" exist, says Hopkins, while the accounts of special creationists affirm that they exist:

[A]ll theories – like those of Lamarck and Mr. Darwin – which assert the derivation of all classes of animals from one origin, do, in fact, deny the existence of natural species at all. (1973, p. 241)

The theory commonly received asserts, on the contrary, the existence of *natural species*, each of which, since by hypothesis they are incapable of being derived from each other, must have had an independent origin. (p. 242)

Darwin could have adopted Hopkins's use and still expressed evolutionism. If he had, he would have concluded his one long argument by saying, "Because complex organisms have been generated by evolution, there are no species."

Hopkins illustrates not only that terminological differences do not force disagreement between Darwin and critics like Hopkins and Wollaston but also that communication can be achieved across the revolutionary divide despite terminological differences. For that reason, the partial communication that Kuhn says is inevitable is less problematic than we might have feared. Practically speaking, it may be inevitable that such miscommunication will often occur, but the miscommunication can be avoided in principle.

## II.3. *Scientific Progress*

Both resistant scientists like Hopkins and revolutionaries like Darwin refine the language to discuss evolutionism. Because both refine the language, a new worry arises – that there is no accumulation of scientific knowledge over time about the same object of study. That 'species' has taken on a new use seems to threaten the idea that science has made progress in the study of species through Darwin's revolution.

Kuhn frequently displays pessimism about progress. He remarks famously that, after a revolution, converts to a new theory no longer address the same *world* they did before: "after a revolution scientists are responding to a different world" (1970a, p. 111). And Kuhn repeatedly casts doubt on the conventional picture of scientific activity as a process of knowledge accumulation. Closely related to his pessimism about whether scientists accumulate knowledge over time about the *same subject of research* is Kuhn's pessimism about whether scientists make progress toward learning *the truth about the world*. Kuhn maintains into his later years the skeptical position that linguistic instability renders "meaningless" the question of "science's zeroing in on, getting closer and closer to, the truth" (Kuhn 2000b, p. 243; see also 1970a).

Kuhn's pessimism concerning whether scientists progress in studying the same subject matter over time is rooted in an important observation, as I have indicated in the previous sections of this chapter. There is good reason to resist, with Kuhn, a naïve picture of knowledge accumulation. The problem with a naïve picture of accumulation is that the subject matter of sentences containing a theoretical term like 'species' changes some over time. When naturalists gradually came to assent to 'Species arise by evolution', they did not merely add this sentence to a stock of sentences about *species* that were accepted by pre-Darwinians. The pre-Darwinians did not even have a term in their language for *species*, the referent of our term 'species'. Their term 'species' was a vague one that did not, given evolution, clearly refer to the groups that arise by evolution. This term, at least as used by those who deferred to common use, did not clearly refer to *anything*, given the changed world view. And as Kuhn would say, "When referential changes of this sort accompany change of law or theory, scientific development cannot be quite cumulative. One cannot get from the old to the new simply by an addition to what was already known" (Kuhn 1981, p. 3).

Still, the threat to science's progress in the study of a single subject (see, e.g., Putnam 1975e, p. 237) is illusory. It is true that scientists' sentences about "species" have changed in meaning, so that the continued accumulation of sentences may not be read as a record of a simple accumulation of knowledge about the same object of study. Even so, this is no worry for the intuitively compelling claim that science progresses. The referent of many a theoretical term changes, but *this reference change occurs precisely because scientists grow in their understanding of* both *the earlier referent* and *the later one.* Darwin understood more than did speakers before him about both what he called "species" and what speakers before him called "species." Similarly, scientists today, at least those who educate themselves about the development of the use of the term 'species', understand more than the scientists of 1850 did about both what scientists today call "species" and also what scientists in 1850 called "species." So even though reference change precludes the accumulation, across the revolutionary divide, of *sentences* that are all about precisely the same pre-revolutionary or post-revolutionary subject, there is still an accumulation of *knowledge* about precisely the same pre-revolutionary subject and about precisely the same post-revolutionary subject. Despite a change in terms' meaning, then, there is a subtle respect in which the subject matter remains constant, if the view I will put forward is right.

This will be my response to the worry that there has been no accumulation of knowledge about species on account of an alleged change of subject. If my response is right, then Kuhn's related skepticism about there being any *truth* about the world can also be resisted. Because scientists study the same entities before and after the Darwinian revolution, there is no worry that there can be no truth about one shared world for scientists' statements to mirror. Linguistic instability of the sort at issue can be acknowledged by those who say that evolutionists are right about the way the world is, that special creationists are wrong about the way the world is, and that therefore the world is such that humans and radishes have a common ancestor.

It is time for more detail. In the following two subsections, I argue that Darwin's revolution reveals two varieties of progress in scientists' successive claims concerning the truth about evolution. The first variety of progress, to be discussed in section (II.3.a), is a straightforward one that I have not yet mentioned. It is manifested in speakers' changed receptivity toward sentences that do *not* shift in meaning as a result of the revolution. The second variety of progress, to be discussed in (II.3.b), is more subtle. It is the type

129

of progress adumbrated above, which occurs precisely when sentences *do* change in meaning – more specifically, when sentences originally expressing vague claims are redeployed to express precise ones.[16]

II.3.a. UNCHANGED PARTS OF THE LANGUAGE. Parts of the language do *not* undergo change through revolution. Some sentences formerly held to be true are later held to be false, and vice-versa, without there having been any change in meaning. This allows for progress as pure and simple as even an old-fashioned logical positivist could hope for. Thus when opponents came to accept the truth of 'Humans have primitive ancestors' or 'Humans and dogs share an ancestor', this marked straightforward progress from error to truth. Neither sentence seems to have altered in meaning during the revolution. The referents of 'human', 'dog', 'ancestor', and so on remain unchanged, allowing for communication across the revolutionary divide.[17]

It does not seem too optimistic to suppose that the entire conflict over evolution could be stated in neutral terms without any appeal to affected terms like 'species' and 'variety'. Robert Chambers, the anonymous author of *Vestiges of the Natural History of Creation*, resorts to neutral language in his sequel, the *Explanations*, as he continues to defend evolution. In this way Chambers circumvents unnecessary, fruitless conflict. After describing several cases of evolution, like that which produced cabbage from a trailing seaside plant, Chambers refrains from announcing that "species" transmute. He is well aware of how opponents would reply:

> It will be said, these changes are all mere variations of specific forms, and the facts do nothing but show that that has been called species which is only variety. But where is this to have its limits? If the cabbage and sea-plant are to be now regarded as one species, it seems to me that we have to go very little further, to come to the lines of successive forms or *stirpes*, which my hypothesis suggests. (1845, pp. 116–17)

How far a "species" can extend given evolution is unclear, but Chambers does not attempt to clarify that issue for the purpose of using the term to frame his own position. Rather, he abandons the term 'species', satisfied to express his view in terms of "successive forms or *stirpes*," a stirp being a branch or line of descent – for example, the birds form one branch and the mammals another (1845, p. 70). By avoiding controverted terms here, Chambers states evolutionism without changing the meanings of the words he uses. So when Chambers's opponents would gradually come to accept the sentence 'The various organisms today belong to stirps of one evolutionary tree', they would not thereby change their use of any terms; rather, they would simply come to accept true statements they formerly rejected. That is straightforward progress.[18]

II.3.b. FROM VAGUENESS TO PRECISION. When language remains stable, the apparent obstacle to progress presented by linguistic change is not an issue. Some sentences remain stable through theory change, or so I have argued. But other sentences, such as 'New species arise by evolution', do not remain stable. When speakers come to accept a sentence like 'New species arise by evolution', they use the sentence to express a *new* claim. Although it is not so obvious that there is progress in speakers' coming to accept a sentence that expresses a new claim like this, there is progress. Progress is made in replacing vague statements that are neither clearly true nor clearly false with straightforwardly true statements. After arguing for this position in section (II.3.b.i) below, I will defend it against the charge that it fails to give scientific accuracy its due in section (II.3.b.ii).

*II.3.b.i. Expressing Scientific Progress with Precision.* Before the Darwinian revolution, 'species' did not refer to *species*: It did not clearly and precisely refer to anything, because the presuppositions for use turned out to be false. Therefore, although the sentence 'New species have arisen by evolution' as *we* use it today is just true, that sentence was *not* just true as earlier speakers used it. It was vague, so the precise status of the sentence's truth value depends upon the preferred account of vagueness: On a standard account, the sentence would be indeterminate, or true to a degree (and false to a degree), or true on some precisifications and not others.[19]

I have argued that there were various ways 'species' might have been made more precise after the Darwinian revolution. All of these ways would more or less have resembled earlier use, but they would also have refined it. The important point here is that *whichever* of these refinements might have been adopted, it would have allowed speakers to express something straightforwardly true or false rather than vague by 'New species arise by evolution' and the like. The move from affirming or denying the vague assertion that pre-Darwinian speakers expressed by 'New species arise by evolution' to correctly affirming the precise assertion that post-Darwinian speakers express by 'New species arise by evolution' represents conceptual progress, despite a change of subject in the relevant sentence.

Take an earlier speaker who denies 'New species arise by evolution' without distinguishing between what we might call a "Darwin-species" and what we might call a "Hopkins-species." This speaker does not quite mean what we do by 'species'. Such a speaker would not feel pressed to distinguish between these two kinds of "species" because, given special creation, every

Darwin-species (the radish, the polar bear, etc.) is a Hopkins-species, which "must by definition have had a separate and independent origin" (Hopkins 1973, p. 241). But given evolution, such a speaker's denial of the sentence 'New species arise by evolution' becomes inexact: New Darwin-species arise by evolution, but new Hopkins-species do not. The speaker has vaguely managed to speak about each and would make progress by distinguishing them. If the speaker retains 'species' for Darwin-species, she will be able to affirm the sentence, and despite a change in what she discusses (she will no longer discuss Hopkins-species even vaguely, and she will definitely rather than vaguely discuss Darwin-species), the speaker will have made progress in her formerly inexact discussion of Darwin-species.

This speaker has progressed in understanding both what she *formerly* called "species" and also what she *now* calls "species." What she formerly called "species" she now understands to be a confusion between Hopkins-species and Darwin-species. She has come to understand that Darwin-species evolve even though Hopkins-species cannot. Hence, there is progress in her understanding of what she *formerly* called "species."

She has also come to understand that what she *presently* calls "species," Darwin-species, are distinct from Hopkins-species, whereas before she ran the two together. And she has come to understand that what she presently calls "species" evolve. She did not understand all of this before. Hence, there is progress in her understanding of what she *now* calls "species," too.

There is progress in her understanding both of what she formerly called "species" and what she now calls "species" and, for that matter, in her understanding of Hopkins-species. This progress occurs alongside a change of subject in sentences employing the term 'species'.

This is my account of progress through linguistic change. In proposing the account, I have said that earlier speakers' statement 'New species have arisen by evolution' was vague. In the following chapter (section I.2), I will say more to defend and clarify my claim that the pre-Darwinian statement is vague. However, the basic line that I take here (and in what follows, through the next chapter) is tenable even if earlier speakers' statement 'New species have arisen by evolution' was *not* vague, so long as earlier speakers' statement suffered from an unfortunate failure to distinguish between Darwin-species and Hopkins-species in a way that later sentences did not, as I have argued. It might be, for example, that the earlier speakers' statement was meaningless, rather than vague, as some have suggested to me.[20] Even if that is right, Darwin made progress: Scientists before him expressed something meaningless by 'New species arise by evolution' because they did not distinguish between Darwin-species and Hopkins-species but rather assigned both to the term 'species'. Later speakers distinguished between Darwin-species and Hopkins-species and thereby came to express something meaningful and true by the sentence. That is progress.

*II.3.b.ii. Giving Accuracy Its Due.* Some commentators[21] have responded to the basic picture of progress above (from section II.3.b.i) with an observation like this: "Scientific progress is not achieved by linguistic precision: Precision without accuracy is not progress." This observation has apparently been thought to undermine my position. It does not.

I agree that precision without accuracy is not progress, but my account of progress does not ignore accuracy in favor of precision. On my account, scientists achieve both greater precision *and* greater accuracy over time, and these two gains are related in an interesting way: Greater precision is desirable *because* it allows speakers to formulate statements that are more accurate. Precision does not *replace* the need for accuracy.

Nor does greater precision *have* to result in a speaker's making more accurate claims. To return to the foregoing example, 'species' was once vague in such a way that 'New species arise by evolution' was not clearly true or clearly false, given evolution. The sentence became clearly true when speakers refined the use of 'species', causing it to refer precisely to Darwin-species. But that change in language did not by itself assure scientific progress or greater accuracy in speakers' statements, because having made the relevant refinement, speakers could have gone on to reject 'New species arise by evolution'. They could have gone on to declare evolutionism false. If they had, their greater precision would have resulted only in their saying something straightforwardly *false*: Speakers would have moved from denying the truth of a sentence whose truth value is indeterminate or unclear because of vagueness to erroneously denying the truth of a sentence that is just true. This is hardly progress.

My claim, then, is only that precision *allows* speakers to affirm something straightforwardly true by 'New species arise by evolution', not that precision forces them to affirm it. Speakers can make the mistake of *denying* the more precise and straightforwardly true claim. If they do, they have lost the opportunity to use their refined language to make more accurate claims and have failed to make scientific progress.[22]

Just as one should not try to contrast my claim that Darwin brings greater precision with the claim that he brings more accuracy, one should not try to contrast my claim that the earlier use of terms like 'species' was vague with the claim that the earlier use of such terms was less scientifically useful than our use. Earlier use was less useful, and it was less useful *because* of the vagueness, which inhibited accurate communication. Of course, this is not to say that the earlier use of the term 'species' to designate what it once designated was *erroneous* or *false*. The decision to use a term to designate one thing rather than another cannot be false. It is the statements that are made with a term on a use that can be false.

## II.4. *Advantages of Moderation*

I have argued that the transition from creationism to Darwinism brings about an interesting conceptual and linguistic transformation that upholds many of Kuhn's observations concerning scientific revolutions. Still, my account is unmistakably moderate, because it acknowledges the theoretical possibility of full communication across the revolutionary divide and because it acknowledges progress in the study of biological kinds and an accumulation of our knowledge about them. Here I draw a close to my discussion of this case after explaining some advantages of my account.

The main advantage of a moderate account like the present one is that it manages to reconcile an apparent conflict between compelling arguments for linguistic change through scientific revolution and compelling arguments for progress through revolution. That there is linguistic change seems hard to deny, and history-oriented philosophers of science like Kuhn are right to emphasize it in their criticisms of earlier philosophers who had emphasized progress. Yet it also seems hard to deny progress, which is not easily accounted for by those who are most zealous to emphasize the dramatic effects of linguistic change. The ideal is to have *both* linguistic change and progress.

Kuhn is himself anxious to preserve progress. Later theorists "preserve a great deal of the most concrete parts of past achievement" (1970a, p. 169), Kuhn grants. Many have questioned whether Kuhn is entitled to recognize enough common ground between a theory and its successor to allow for this kind of preservation (see, e.g., Salmon et al. 1992, pp. 149–50).[23] It seems clear that the moderate account offered here (section II) is not problematic in this way. It leaves enough common ground that a later theory could preserve much from an earlier one, either in the form of statements that do not lose their original meaning or in the form of refined descendants of vague statements.

Just as later theories preserve some claims of earlier theories, they also jettison some claims. There is disagreement between the theories, and this too requires common conceptual ground. A more radical account of theory change than the moderate one offered here runs into difficulties accounting for disagreement.[24] My own account easily allows for speakers past and present to disagree about many statements expressed in neutral language – for example, 'Darwin and Rover share an ancestor'. Speakers past and present also disagree in a more subtle way about the truth of sentences in which terms have changed meaning. For a special creationist who fails to distinguish between Darwin-species and Hopkins-species, 'New species evolve' would be false, because on her account neither Darwin-species nor Hopkins-species evolve. But in the former stance, which she does not distinguish from the latter, she disagrees with Darwin.

By acknowledging linguistic change and then examining the apparent threats thereby introduced, the present account also helps to illuminate the reasons skepticism has seemed plausible to many. It is important in this way to account for the evidence on which skepticism rests. As Kitcher remarks, "sceptical positions cannot be satisfactorily dismissed with a quick *reductio*" (1978, p. 546n.). The present account provides an explanation of what is right about skeptics' accounts and where they go wrong, rather than the common but less helpful observation that something must be amiss with a theory that fails to recognize scientific progress.

### III.  A SECOND CASE: THE OVERTHROW OF VITALISM

Not every theoretical advance has affected terms in the way Darwin's did. Nearly two millennia ago, Galen criticized fellow biological theorists for denying that urine moves from the kidneys to the bladder through the ducts that run from each kidney to the bladder. To refute his opponents, Galen tied shut the ducts, known as 'ureters', in a pig. Urine backed up in the bloated ureters behind the blockages and failed to enter the bladder until the ties were removed. This and similar experiments provided convincing evidence that the ureters do carry urine. But such a discovery hardly changed the meaning of the relevant theoretical terms 'duct', 'urine', 'bladder', and so on (in Greek). Galen's opponents used these terms just as Galen did when they denied that urine passes through the ureters. They were just wrong about how urine gets to the bladder.

In the same way, Antonie van Leeuwenhoek's critics were wrong in the seventeenth century to think that he had not seen tiny living creatures through his microscope. Leeuwenhoek had indeed seen and described bacteria and protozoa. Critics did not use the expression 'living creature' differently than Leeuwenhoek, when they accused him of inventing "fairy tales." They used the expression as he did. They were just wrong about whether Leeuwenhoek saw living creatures: Some, for example, objected that had Leeuwenhoek

135

boiled his liquid, he would still have seen all of the very same particles moving about in the same way.

Similarly, to take another couple of celebrated cases from the seventeenth century, Francesco Redi was right to suppose that maggots are not generated spontaneously from meat; they come from eggs laid by flies. His demonstration that this is the case did not change the meaning of 'maggots', 'meat', 'flies', or 'eggs'. And William Harvey did not change the meaning of 'blood', 'heart', 'veins', 'arteries', or 'circulates' when he demonstrated that blood circulates to and from the heart through the veins and arteries. This was not the received view. But Harvey's opponents were wrong, and he was right: Blood *does* circulate.

There are many similar cases of scientific change that seem not to have resulted in a warping of the language in the way that the discovery of evolution did. But although many advances seem to leave untouched the terms used before the advance, Darwin's advance is not uncharacteristic in illustrating linguistic change. Another example of a change that is in important respects revolutionary will help to broaden and deepen the present investigation by emphasizing somewhat different Kuhnian observations. A brief investigation of the doctrine of vitalism and its decline will serve this purpose. It will serve to illuminate Kuhn's attack on the rewriting of history.

### III.1. *Vitalism as It Is Understood Today*

Vitalism is of unquestioned historical significance, boasting an unusually distinguished list of adherents from Harvey to Louis Pasteur and Hans Driesch. As one source says, "The roll call of vitalists includes most of the classical figures in the history of the life sciences" (Bynum 1981, p. 440). Espoused by central thinkers, vitalism reached into the center of biological theorizing. There at the center of biological discourse, vitalistic doctrine proved fertile: It inspired groundbreaking experiments. Pasteur's famous experiments on spontaneous generation and on fermentation, for example, were driven by vitalistic hypotheses.

Indeed it has been argued that belief in vitalism was necessary to get biology established as an independent science. François Jacob writes that vitalism is a theory that early biologists "had to postulate in order to acquire independence" (1982, p. 244). Vitalists maintained that the living world is in some important respect special. It cannot be reduced to the sorts of phenomena that physicists and chemists study. Typically, vitalists would refer instead to a "vital force" that was supposed to distinguish living from inanimate matter. Those opposed to vitalism, generally known as mechanists, insisted, on the other hand, that biological phenomena reduce to chemical and physical phenomena. Unfortunately, as Mayr points out, early mechanists drastically oversimplified their biological subject matter as they attempted to ape physics and chemistry. This hampered their productivity (Mayr 1982, p. 96). Many authors have noted the greater success of vitalists (Mayr 1982, p. 848; Bynum 1981, p. 440).

In spite of its illustrious history, vitalism became moribund by the end of the nineteenth century and was unanimously rejected after the deaths of a few holdovers into the twentieth century, most notably Driesch. No biologist today professes belief in vitalism. It is natural to suppose that serious problems must have been responsible for the routing of such a well-entrenched belief. Certainly standard textbook treatments urge just that. But a Kuhnian skeptic will suspect that such treatments reflect a rewriting of history by historical winners.[25] The result is Whig history, a picture of progress tending "to make the history of science look linear or cumulative" (Kuhn 1970a, p. 139) at the expense of misrepresenting vitalists.

There is considerable merit in the Kuhnian position. In general vitalism is today summed up by two related positions. The first of these positions is that living organisms are powered in a way that nonliving objects like rivers and volcanoes are not, by an immaterial substance, rather than by a particular arrangement of physical and chemical constituents. Thus, Medawar and Medawar (1983, pp. 275–6) characterize vitalism as

> a doctrine that takes various forms, which have in common a flat repudiation of the idea that a living organism's vivacity – its state of being alive – can be explained satisfactorily in terms of its form and composition; that is, in terms of what it is made of and how those constituents interact physically and chemically. Some immaterial vital principle is required in addition.[26]

The second position associated with vitalism is that only a living organism can synthesize organic chemicals. This characterization of vitalism may be found in countless textbooks (see Cohen and Cohen 1996 for many references) as well as popular accounts (e.g., Williams 1995) and histories (e.g., Mayr 1982, p. 52).

These two doctrines, which are generally taken today to sum up vitalism, may be labeled "the thesis of substantial vitalism" and "the thesis of chemical vitalism," respectively. Neither doctrine is seriously entertained in biological work today. So it is not surprising that Mayr should insist that "Vitalism in all of its forms has been totally refuted and has had no serious adherency for several generations" (1982, p. 131).[27]

## III.2. *The "Vitalism" of Past Speakers*

Is it true that "vitalism in all its forms" is in such a neglected, defeated state? Certainly not all forms of what *self-proclaimed* vitalists knew as "vitalism" are in such a dismal state. Neither substantial vitalism nor chemical vitalism has been accepted by all "vitalists." Driesch's version of vitalism, for example, rejects chemical vitalism. Some "vitalists," like the great chemist Justus von Liebig, have rejected both substantial vitalism *and* chemical vitalism. Liebig was a pioneer in organic chemistry and so he naturally rejected chemical vitalism (see, e.g., Lipman 1967, pp. 168ff.). He also rejected substantial vitalism, writing, "There is nothing to prevent us from considering the vital force as a particular property, which is possessed by certain material bodies, and becomes sensible when their elementary particles are combined in a certain arrangement or form" (Liebig 1842, p. 198).

Some versions of "vitalism" are held by many scientists today, though no longer under the rubric 'vitalism'. Consider, for example, the "vitalism" of the great experimentalist Claude Bernard. Bernard repudiated cruder versions of vitalism, but he also objected to "chemists and physicists who . . . try to absorb physiology and reduce it to simple physico-chemical phenomena." He settled on a position he would call "physical vitalism," which recognizes that "biology has its own problem and its definite point of view" but that also recognizes physico-chemical determinism.[28]

Bernard's position that biological concepts and theories do not reduce to physical ones even though biological processes are physico-chemical ones is not well described as having "had no serious adherency for several genera-tions" (Mayr 1982, p. 131). On the contrary, it is a position held by Mayr him-self (1982, pp. 60–3). Of course Mayr would not call this *anti-reductionism* "vitalism" of any shape or form. He explicitly denies that anti-reductionism is a version of vitalism (1982, p. 59). But here he does not use 'vitalism' in the same way Bernard does.

What makes Bernard's use of 'vitalism' for his anti-reductionism appro-priate is that anti-reductionism is one of a family or cluster of positions in the tradition of "vitalism." A number of nineteenth-century thinkers defended vitalism precisely on the grounds that biology is not just applied physics and chemistry (see, on this point, Gregory 1977, p. 166), while mechanists at-tacked vitalism on the grounds that physiology would dissolve into physics and chemistry (see, e.g., Coleman 1977, p. 151).

The qualifier 'physical' in 'physical vitalism' emphasizes the difference between Bernard's variety of "vitalism" and others, some of which in-clude less judicious selections from the cluster of positions associated with 'vitalism'. Bernard is not alone in accepting a form of "vitalism" like the one that he adopts. E. S. Russell (1911), for example, also rejects substantial and chemical forms of vitalism but, like Bernard, opposes the reductionistic views of mechanists, endorsing a position he later calls "methodological vitalism" (1930, p. 177).

Anti-reductionism was once, but is no longer, associated with the term 'vitalism'. Another position once associated with the term 'vitalism' is that life is an *emergent* property. An emergent property is roughly a property of matter that is in some important sense new and unpredictable from knowledge of the matter's constituents (Morgan 1923). Consider one of the qualities of water: translucence. No one could have predicted water's translucence simply by experimenting with hydrogen alone and oxygen alone, to learn their properties. Water introduces new properties that emerge at a higher level than the level of hydrogen atoms and oxygen atoms.

Vitalists embraced emergence in the organic world (see, e.g., Emmeche, Køppe and Stjernfelt 1997, p. 86). At the level of life, they insisted, new properties emerge. Therefore, it was natural that C. D. Broad, for example, who accepted emergence at the level of life even though he rejected the "Substantial Vitalism" of Driesch, should have called his position "Emergent Vitalism," to stress his continuity with vitalists and differences with mechanists (Broad 1925, pp. 58, 67–9). 'Emergent vitalism' is a rubric used by many others, as well, up until almost the middle of the twentieth century (McDougall 1934, p. 108; Wheeler 1939, pp. 170ff.; see also Russell 1930, p. 179).

Today few associate the label 'vitalism' with emergence of any kind: The former is taboo, and the latter is widely accepted. Thus, Medawar and Medawar write that "vitalism is in the limbo of that which is disregarded." Emergence, on the other hand, is a valued concept. "The state of being alive seems to us to be best described as an emergent property," they affirm (1983, p. 277). Similarly, Mayr enthusiastically embraces emergence but rebuffs the idea that this is vitalistic: Modern emergentists "accept constitutive reduction without reservation and are therefore, by definition, nonvitalists" (1982, p. 64). Constitutive reductionism is just the doctrine that organisms are composed of chemicals found in the inorganic world (Mayr 1982, p. 60). So by saying that constitutive reductionists are nonvitalists by definition, Mayr is in effect insisting that in order to be a vitalist one must subscribe to Broad's "Substantial Vitalism," which is rejected by virtually everyone today and which was rejected by Broad.

Mayr wants to keep the unattractive word 'vitalism' away from any useful notion. But he is certainly not using the term as Broad and others who recognized "Emergent Vitalism" did. He is using the term in a new way. To be sure,

Mayr is using the term as writers now generally do. He is offering the sort of definition that a standard reference source would (see, e.g., Beckner 1967, p. 254). But that is just because 'vitalism' does not have the same meaning now as it did closer to the controversy.

### III.3. *Chance and the Inevitability of Scientific Conclusions*

Many of the theses that earlier thinkers counted as "vitalistic" are still widely held. This suggests that the current scientific conclusion that vitalism is unacceptable was not inevitable; good scientists could have concluded otherwise, and there would have been no loss in doing so.

The suggestion that today's scientific conclusions were not inevitable is a popular one among disciples of Kuhn, who insist that science is a social construction. There is something to such a claim. Culture may affect whether scientists understand or accept a theory, as sensitive philosophers who insist on science's objectivity have no trouble agreeing (Ruse 1999; Kitcher 1993, pp. 11ff.). Critics of scientific rationality want to go further than this, holding, as Kitcher relates, "that those who resisted past decisions that we endorse as 'correct' may have been just as reasonable as their opponents" and even that "our endorsement itself is crucially dependent on that initial decision" (1993, p. 7; see also Collins 1985, pp. 131–5; Kuhn 1970a, pp. 3–5; Rorty 1999, pp. 121–2).

There is a *respect* in which this is accurate. Earlier speakers used the term 'vitalism' to cover a welter of views that were often enough not distinguished from one another. 'Mechanism' and 'vitalism' have been afflicted with immense vagueness (McDougall 1938, p. 75). As Broad says, "both names cover a multitude of theories which the protagonists have never clearly distinguished and put clearly before themselves" (1925, p. 44). Given the unclarity of use, and given that some positions on the vitalists' side, such as emergence and anti-reductionism, are popular to this day, 'vitalism' could have continued to be associated with these popular theses rather than with now-discarded theses, as it has been. Had that happened, people today like Mayr would be calling themselves "vitalists" and assenting to sentences like 'Vitalism is scientifically acceptable'.

Does all of this suggest that vitalism (i.e., what *we* today call 'vitalism') is, after all, viable and that scientific conclusions are up for grabs? No. Some theses associated with the word 'vitalism' by *earlier* speakers are viable, but that is a different matter. Different versions of "vitalism" embraced tenable or untenable theses associated with the word, or some combination of tenable and untenable theses. Those versions of "vitalism" that embraced just tenable theses were clearly right. Those that embraced just untenable theses were clearly wrong. Some other versions had some matters right and others wrong.

In the same way, whether theses called 'vitalistic' *today* are viable is not somehow up for grabs. The thesis of chemical vitalism, for example, is just untenable. Vitamin C can be produced in a lab as well as on a citrus tree. It is simply false to suppose, as some vitalists of centuries past did, that an organism is needed to produce organic chemicals. This thesis of chemical vitalism must be rejected by rational people with the available knowledge.

### III.4. *Politics and Misrepresentation*

Because chemical vitalism has been refuted so soundly, and because the complicated details of conceptual confusion and development are unnecessary for chemistry students, it is not surprising that chemistry textbooks should summarize vitalism as the dead thesis that organic chemicals must be produced in an organism. As Kuhn says,

> For reasons that are both obvious and highly functional, science textbooks (and too many of the older histories of science) refer only to that part of the work of past scientists that can easily be viewed as contributions to the statement and solution of the texts' paradigm problems. (1970a, p. 138)

History is rewritten by later generations. Still, it is hard to see that historians *need* to oversimplify in order to convey progress, as Kuhn sometimes suggests. It may be true that for today's writers "the outcome of revolution must be progress, and they are in an excellent position to make certain that future members of their community will see past history in the same way" (1970a, p. 166). But progress is visible even when history gives earlier thinkers their due. Mayr would do as well to convey progress even if he came to see that he has more in common with Bernard, an acknowledged vitalist (Mayr 1982, p. 106), than with the mechanists he lists as forebears (p. 114). If he decided to retract the label 'mechanistic' from his own views (p. 59), nothing would be lost. He could still say that substantial and chemical vitalism have been refuted or at least found unfruitful, and count their passing as progress.

In the same way, biographers and eulogists sometimes needlessly suppress embarrassing information, leaving this information for social constructionists to ferret out. Liebig's biographers have tended to downplay his vitalism (see Lipman 1967, pp. 175–6). Bernard's pupils left telling omissions in an otherwise very complete index in order not to draw attention to his "vitalism" (Olmsted and Olmsted 1952, p. 147). Such selective representation may exaggerate achievement, but it is hardly needed to convey achievement. Great figures can take some positions that are later found wrong or partly wrong or unfruitful. And great figures can certainly employ terminology that goes out of fashion to express ideas that are more lasting.

Sometimes the politics of conceptual change are less benign than in the foregoing cases. Conceptual change can result in the martyrdom of scientists slow to make adjustments. Schultz (1998) describes, for instance, the discovery in the late nineteenth century that living frog skin can actively absorb, as well as secrete, fluids. Because dead skin does not have the same powers, the discoverer, E. Waymouth Reid, announced in 1892 that he had found the presence of "a vital absorptive force" in living frogs' skin. Schultz remarks that the discovery marked a conceptual breakthrough in epithelial biology that might have been found in a reputable journal even in the middle of the twentieth century. But Reid's observations went ignored after he published them, and his use of vitalistic terminology may well have been the reason. "While it is not clear whether E. Waymouth Reid was an 'orthodox vitalist' or whether he used the phrase 'vital absorptive forces' loosely, it seems that his thesis never had a chance!" (Schultz 1998, p. C14).

Similar stories abound.[29] But despite some radicals' suggestive discussions, such stories do not show that might makes right, or that anything goes in good methodology, or that truth is a matter of vote, or anything else that undermines scientific objectivity. Reid was *right* that living skin can absorb fluids in a way that is comparable to secretion. If Reid's observation went unnoticed or undervalued because of vitalistic beliefs or terminology, that is another matter. Not only was Reid right about living skin's power of absorption, but nonvitalists can *recognize* that he was right. Indeed, it is a present-day scientist (Schultz) who calls our attention to the correctness of Reid's insight, more than a century later.


IV.   THE FATE OF INCOMMENSURABILITY

I have argued that scientific revolutions are accompanied by linguistic change, and that this linguistic change supports important Kuhnian observations about revolutionary transitions in science. At the same time I have argued that none of this compromises the intuitive claim that scientific advancement is progressive. I have said little about whether there is *incommensurability* between rival theories. Part of the reason for this omission is that there is much disagreement over what incommensurability is, as I will explain.[30] Whether there is any incommensurability depends, of course, on what incommensurability is.

### IV.1. *A Strong Interpretation*

According to some authors, incommensurability is strict incomparability, which is alleged to hold between different theories as a result of linguistic change. Feyerabend, for example, suggests this interpretation of 'incommensurability' when he says of incommensurable theories, "Their *contents* cannot be compared" (1988, p. 226). The interpretation of 'incommensurable' as *incomparable* accounts well for passages like this one, and many authors have taken this interpretation.[31]

If 'incommensurable' is interpreted to mean *rationally incomparable*, then my earlier arguments suggest that the incommensurability thesis has little to recommend it. The linguistic change that accompanies revolutions can be accepted with equanimity by those who reject this kind of incommensurability. Nothing about such linguistic change calls for any conclusion to the effect that rival theories are incomparable in any respect that threatens the rational progress of science.

### IV.2. *A Weak Interpretation*

On other interpretations, incommensurability is not the incomparability alleged to result from linguistic transformation. Rather, incommensurability just *is* the linguistic and conceptual transformation that accompanies scientific revolutions. Douven and Van Brakel, for instance, say that semantic incommensurability is roughly the doctrine "that scientific revolutions induce referential shifts so that scientists before and after a revolution are referring to different entities, even when using the same words" (1998, p. 60); others have recommended a similar use for the term (e.g., Sankey 1994, p. 221).

If incommensurability just is the linguistic and conceptual transformation that Kuhn and others find in scientific revolutions, then the upshot of my own argument is that different theories *are* incommensurable. I have argued that such change must be recognized. But if this is what 'incommensurability' stands for, the radical baggage associated with the word must be jettisoned. Incommensurable theories may be compared rationally, and the succession of incommensurable theories marks progress.

### IV.3. *A Moderate Interpretation*

There is another common interpretation of 'incommensurability'. On that interpretation, 'incommensurability' signifies untranslatability. Although in earlier writings (Kuhn 1970a, pp. 202ff.; cf. 2000b, p. 238) Kuhn says that

143

there can be translation between incommensurable theories, he takes incommensurability to be untranslatability in later writings: "The claim that two theories are incommensurable is then the claim that there is no language, neutral or otherwise, into which both theories, conceived as sets of sentences, can be translated without residue or loss" (Kuhn 1983a, p. 670; see also 1979, p. 416). In another place he writes more tersely: "Incommensurability thus equals untranslatability" (1990, p. 299), albeit a restricted type of untranslatability. The claim that rival theories are not intertranslatable may appear to raise the specter of incomparability again. A brief investigation of the thesis will quell this fear. There are two rough, prominent arguments for untranslatability.

IV.3.a. DUBBING CEREMONIES PRESUPPOSE THEORY. One argument, skillfully articulated in an illuminating discussion by Sankey (1994, Chapter 3), is that the terms of a theory may be coined in a way that presuppose the theory. This might forbid the creation of synonyms in the languages of rival theorists. Feyerabend suggests this when he says that

> mere difference of concepts does not suffice to make theories incommensurable in my sense. The situation must be rigged in such a way that the conditions of concept formation in one theory forbid the formation of the basic concepts of the other. . . . (Feyerabend 1978, p. 68 note 118)

Thus, the Aristotelian term 'impetus' may be indefinable on Newton's Theory, because it is supposed to designate a force acting upon all projectiles, but for Newton there is no such force acting upon a projectile in a state of inertial motion (Feyerabend 1981a, pp. 62ff.).

Proponents of one theory may also find themselves unable to make use of one or more of *several* ways of establishing a term's reference, on such an account. Special creationists could coin their term 'species' both by appeal to examples, like the *giraffe* or the *radish* and *also* by a baptism in which speakers say something like "A species is any group like *that* and all of its blood relatives" (the speaker points to a giraffe). A Darwinian, who does not recognize original pairs, would be prevented from the latter course, so his term would not share the meaning of his rivals'. Or so the argument goes.

But the argument is not convincing. It is true that the proponents of a theory may be unable to coin terms for a rival theory *that fully refer* to something according to *their* preferred theory. The terms would be empty or partially referring. But they could still be coined. Darwin *could* have coined 'species' thus: "By 'species' I mean any group including a specially created original pair and all of its descendants." It is just that Darwin would then have had to assent to 'Species do not exist'. This is what Hopkins told Darwin he ought to have said.

Terms for entities whose existence is *uncertain* may be coined in the same way, of course, and it could obviously be useful to coin them. Europeans who first stumbled upon the platypus were skeptical. Taverns and inns commonly displayed stuffed "mermaids" and other monsters that were actually cobbled together from the remains of several organisms. Such macabre displays inclined some naturalists to suspect that the platypus was a product of surgery (Ritvo 1993, pp. 249, 254 note 42). But that need not have prevented them from coining 'platypus' something like as follows: "By 'platypus' I mean the species instantiated in the organism before me, if the organism is the original owner of all of its parts, and if not then 'platypus' is not to refer." No faith in the existence of a platypus species is needed for this kind of dubbing ceremony. The dubber would be free to affirm or deny the sentence 'The platypus species exists'.[32]

IV.3.b.  KUHNIAN LOCAL HOLISM. It is unlikely, then, that holding one theory could prevent a theorist from coining terms used in rival theories. Such coining would be a form of translation: Rival theorists each have the resources to coin words for the same entities.

But Kuhn raises another objection to the coining of terms used in a rival theory. For Kuhn, the members of a language community must share a cluster of related concepts. The untranslatability obtains between ordinary natural languages as well as scientific languages. An example from an ordinary language is provided by the French term 'doux'/'douce', which means something like *sweet*, but which acquires significance by comparison with other terms. 'Doux'/'douce' and other such terms

> belong to clusters of interrelated terms a number of which must be learned together and which, when learned, give a structure to some portion of the world of experience different from the one familiar to contemporary English speakers. Such words illustrate incommensurability between natural languages. In the case of '*doux*'/'*douce*' the cluster includes for example, 'mou'/'molle', a word closer than 'doux'/'douce' to English 'soft', but which applies also to warm damp weather. (1983a, p. 680)

Because "the meaning of 'doux' consists simply of its structural relation to other terms of the network," there can be no English translation. English has the word 'sweet', and this term has analogous relations to terms like 'soft' and 'sugary', but the distances between the terms of the cluster are not quite the same. So in translating 'doux' it will often be proper to use 'sweet', but sometimes it will be better to use 'bland', as for unseasoned soup, or 'soft', and so on. Because the similarity in these uses is captured in the French, with its single term 'doux', but not in English, none of these translations is perfect (Kuhn 1983a, p. 680). So the terms do not quite translate.

Let it be granted that 'doux' cannot be translated perfectly for these reasons. Is this concession worrisome? No. The real interest in translation derives from an interest in making sure that the statements of one language have the right consequences for statements made in other languages. To use a Wittgensteinean metaphor, the logical gears of statements in two theoretical languages should mesh.[33] And even if 'doux' is untranslatable, English speakers can reason about the state of being doux *by learning some French*. Anyone who knows how to use the word can tell whether a particular dish discussed in English *or* French should be called "doux" or whether it should not, whether the matter is unclear, and so on. If it were not possible to reason in English with a word that has no perfect translation in English, then often even reasoning in English *without* the borrowing of foreign words would be impossible. The English word 'hard' has no perfect translation in *English*, so it is like the borrowed 'doux' in that respect. If we had to do without the word, we would use the translation 'difficult to achieve' in some contexts and 'firm' in others. Yet no one proposes that this is any barrier to reasoning with the term 'hard'.

We can reason with terms from other languages. So, of course, can scientists and historians. Alternatively, scholars might choose, rather than to use the terms of a foreign language, to enrich their *own* language by adding terms to it, like 'vital force' or 'phlogiston', or what-have-you.[34]

Finally, without translation, borrowing, *or* enrichment, interpreters may *explain* a text. Translation, in the strict sense Kuhn understands it, is unnecessary for reasoning about statements made in two languages, and so is the use of older terms. It is enough to *mention* the older terms and talk about their conditions of application. This would not be to translate (Kuhn 1983a, p. 672), but it may be used instead of translation to convey an idea.

I conclude, therefore, that if 'incommensurability' means *untranslatability*, then even if rival theories are incommensurable, this does not jeopardize the possibility of nondistorted reasoning between proponents of competing theories.

## V.   MEANING CHANGE AND THEORY CHANGE

In conclusion, the causal theory of reference does not block the worry of incommensurability. Fortunately, the worry is unfounded. Scientific change is accompanied by linguistic change, but this change does not undermine rational progress. That is my response to the incommensurability thesis.

This response to the apparent threat of incommensurability preserves the intuitive conclusion that scientific change is progressive. However, the arguments in favor of this response depend upon considerations that run against another doctrine that many, following Quine, find intuitive: namely, the doctrine that there is no distinction between a change of theory and a change of meaning.

I have argued that sometimes there is theory change without meaning change. Sometimes conceptual change is not like that which attended 'species' or 'vitalism'. Galen, for instance, demonstrated that the ureters carry urine from the kidneys to the bladder. He did not change the meaning of his words for 'ureter' or 'bladder' or 'urine'. Harvey demonstrated that blood circulates. He did not change the meaning of 'blood' or 'circulates'. His opponents, some of whom thought that the arteries carry not blood but rather an air-like substance called "pneuma," were just wrong. Redi did not change the meaning of 'maggot' or 'meat' or other theoretical terms when he demonstrated that maggots are not generated spontaneously from meat but rather that maggots are generated only from eggs laid in the meat by flies. Redi just showed to the satisfaction of the general community that meat not exposed to flies does not generate maggots, even if it has all of the other important conditions for the production of maggots. In all of these cases, the experimentalist faced opposition, and the opposition just lost. Key terms retained their meanings through theory change in a way that they often will not.

Continuity of reference like that in the foregoing examples seems impossible on the view that there is no difference between meaning change and theory change. By changing our views on the spontaneous generation of flies from meat, we ipso facto change the meanings of our theoretical terms, if meaning change cannot be distinguished from theory change. In that case, earlier speakers like Redi's opponents did not assert quite what Redi denied. Redi could not, in that case, have found them to be flat wrong, contrary to a natural account of the case. For in coming to their new beliefs, scientists would have changed the meaning of key terms used to express those beliefs. This conclusion seems to run afoul of any number of advances in science.

Just as theory can change without meaning, meaning can change without theory, I have argued. Recall that our current use of 'species' differs somewhat from that of previous speakers. Earlier users of 'species' assumed the special creation of each "species." When convinced of evolution, earlier speakers might have refined the use of 'species' in various ways. I have argued that any of these altered uses of 'species' would allow for the articulation of both evolutionism and special creationism. But this seems to presuppose a distinction between meaning change and theory change, for if we change the meaning of a term we thereby automatically change the theory at issue, if meaning change and theory change are not distinguishable.

The point can be made clearer if we consider two communities who come to use 'species' differently. Hopkins, recall, says,

> Every natural species must by definition have had a separate and independent origin, so that all theories – like those of Lamarck and Mr. Darwin – which assert the derivation of all classes of animals from one origin, do, in fact, deny the existence of natural species at all.

Suppose that Hopkins and other literate individuals were to retain this use of 'species' in a certain community. In some other community located at some other place, suppose that Darwin and others were to use 'species' as we in fact use the term now. In this way, two different dialects develop: The communities use 'species' differently. Suppose that both language communities eventually concede the evolution of higher organisms from primitive ones. After this eventual concession, the Hopkinsian-language community would have to say, "Species do not exist," because as they use 'species' it cannot refer, in the event that evolutionism is true. Hopkinsian speakers would have to use another word for those evolving entities that *we* call "species" – for example, they might call them "basal lineages." The Darwinian-language community, on the other hand, would say things like "Species exist, and they evolve."

The two communities agree on all placements in the evolutionary tree, we may suppose – for example, that chimpanzees and humans have a closest common ancestor to the exclusion of all other extant primates. Their diagrams of the tree of life are identical. In fact, we may suppose that each group fully accepts all of the received biological statements of the other group, after terminological clarification: The Darwinian community says, "When the Hopkinsians refer to 'basal lineages' they mean *species*," and the Hopkinsian community says, "When the Darwinians refer to 'species' they mean *basal lineages*." Here it seems clear that the communities would fully agree with each other about all of the *facts*. The difference is purely one of language.

Such a conclusion seems not to be permitted on a Quinean view of language change. On such a view, because there is no *difference* between a change of language and a change of theory, the two communities must have a difference of theory by virtue of their linguistic differences. This is not an appealing conclusion.

Considerations like the foregoing cast doubt on the popular doctrine that theory change and meaning change are indistinguishable. My emphasis on *conceptual revision* that is both theoretical and linguistic, on the other hand, may appear to pull in a *different* direction. The kind of conceptual upheaval that I have examined in the past three chapters may appear to *blur* the distinction between theory change and meaning change. It is time to take a closer look at such revision in connection with the Quinean doctrine that theory change and meaning change are inseparable.

# 6

# Meaning Change, Theory Change, and Analyticity

In the past few chapters I have emphasized that conceptual revision is often a blend of theory change and meaning change. Consideration of such revision has moved many philosophers to say, with W. V. Quine, that there is no *difference* between a change in theory and a change in meaning. Quine argues that there is no such difference in the course of attacking analyticity. I examine Quineanism in the present chapter.

Quineans are right to emphasize the importance of conceptual revision that is a combination of theory change and meaning change. But does this sort of conceptual revision show that there is no distinction between theory change and meaning change? I argue that it does not. In section (I), I argue that the relevant kind of conceptual revision provides no convincing grounds for rejecting a distinction between meaning change and theory change.

In section (II) I argue that most philosophers are committed to *accepting* the distinction between meaning change and theory change, because they are committed to analyticity. Kripke has convinced the generality of philosophers that there are necessarily true identity statements, as I have pointed out in Chapter 2. I argue in section (II) of the present chapter that those who accept Kripke's arguments for the necessity of identity statements associated with the causal theory of reference have committed themselves to analyticity. The commitment seems to have gone unnoticed. Indeed, philosophers have been encouraged by Kripke's work to dispense with analyticity. They have supposed that Kripke shows how to accept necessity without analyticity. Nevertheless, if Kripke is right about necessity, as the dominant tradition holds, then Quine cannot be right about analyticity, or about the lack of a distinction between theory change and meaning change.

## I.  CONCEPTUAL REVISION

Why has the Quinean thesis convinced so many? The reason is evidently not that philosophers have been convinced by Quine's argument for the indeterminacy of translation.[1] On the contrary, philosophers have largely rejected the thesis of the indeterminacy of translation, and Quine's related conclusion that meanings are to be rejected altogether. Instead, as I suggest above, philosophers seem to have been convinced by considering *conceptual revision* that meaning change is inseparable from theory change.

### I.1.  *The Quinean Interpretation of Conceptual Revision*

The most popular examples cited in support of Quine are probably provided by geometry. Consider how our understanding of triangles has changed, for instance. In earlier centuries physicists, mathematicians, and educated people in general considered it an obvious conceptual truth that "The sum of the angles of any triangle add up to 180°."[2] We now reject this claim on the grounds that in non-Euclidean geometries, things turn out otherwise. In plane elliptic geometry the angles of a triangle add to more than 180°. In hyperbolic geometry, the angles sum to less than 180°. Something that seemed absolutely certain to earlier thinkers has been rejected. This change, the Quinean maintains, illustrates what we might call "theorymeaning change," theory change and meaning change being indistinguishable. The view that "there is no real distinction between a change of language and a change of view" (Harman 1973, p. 106) enjoys something close to the status of orthodoxy today.[3]

I will argue that Quineans are right in saying that there has been both theory change and meaning change, before I turn to dispute theorymeaning change. That there has been a change of theory is probably most evident, because according to general relativity, real *physical space* is not Euclidean. Rigid rods arranged into a triangular formation could exemplify in our own physical space a triangle whose angles do not add to 180°.

A corresponding revolution in pure mathematics occurred earlier than the advent of the theory of general relativity. In the nineteenth century, mathematicians found out that there are consistent non-Euclidean geometries. It was after this discovery that the question of whether physical space is Euclidean dawned. Gauss, one of the discoverers of hyperbolic geometry, attempted to test the presumed Euclidean character of physical space before general relativity was conceived. To measure the angles of a large triangle, he stationed

151

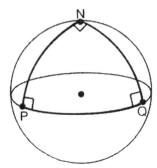

**Figure 6.1.** Adapted from E. Maor, *To Infinity and Beyond*. Copyright 1987 by Birkhäuser Boston and reprinted in 1991 by Princeton University Press. Used with permission of Springer-Verlag.

surveyors on three mountain peaks, though the scale of the triangle turned out to be too small to reveal any interesting results.[4]

That there has been theory change about what earlier speakers would have called "triangles" seems undeniable. That there has been *meaning* change is initially much less apparent. Yet there has been meaning change. Consider, for example, the sentence 'The sum of the angles of a triangle may add up to 270°'. In plane elliptic geometry, triangles may have angles summing to 270°, so present-day mathematicians will affirm the sentence. Earlier mathematicians would have rejected the sentence. Still, we do not quite address the same proposition that they would have addressed, because we do not quite mean the same thing by 'triangle'.

To see how a triangle could have angles summing to 270°, consider a triangle drawn on a sphere. The sphere is a model for plane elliptic geometry. Triangle NPQ in Figure 6.1 has three right angles. Is this a counterexample to the ancient doctrine that a "triangle" must have angles summing to two right angles? Not quite. Earlier speakers would not have recognized it as an instance of what they were talking about by 'triangle' in the first place. So it seems dubious that they straightforwardly did refer to this shape by 'triangle' and that that is that.

The term 'triangle' had its meaning determined in part by *paradigmatic samples*. Speakers gave meaning to 'triangle' in part by referring to illustrations and saying things like "This figure is a triangle," and "That figure is not a triangle proper: It is a so-called 'spherical triangle'." Earlier speakers, who were well aware of spherical triangles like NPQ, and who even called them "spherical triangles," would have objected to the idea that such a figure is a genuine counterexample to their rule that "A triangle's angles must sum to

152

two right angles." Spherical triangles, they would have objected, are not real triangles, strictly speaking.

That meaning has changed becomes still more apparent when we consider that a triangle must be formed by three straight line segments, but the meaning of 'straight' changes when Euclidean geometry is left behind. For Euclid, a straight line segment could be extended continuously, as he affirms in his second postulate. Before the nineteenth century, this was implicitly understood to be the idea that a straight line has infinite length. In the nineteenth century, Bernhard Riemann reinterpreted Euclid's second postulate, assuming that straight lines are *unbounded* but not infinite. That is, one never reaches the end of a straight line even though it circles back on itself. This allowed Riemann to count the great circles of a sphere "straight." A great circle around a sphere is a circle whose plane passes through the sphere's center. So in Figure 6.1, the great circle passing from P to Q and back around to P is a "straight" line according to Riemann's reinterpretation of Euclid's second postulate.

Elliptic geometry goes further in departing from the Euclidean account of "straightness." It violates Euclid's first postulate, on its standard interpretation. According to the first postulate, there is just one straight line between any two points. But if a great circle is straight, then it is clear that there is more than one straight line between any given pair of points, and one line may be much longer than the other! Between P and Q are two distinct straight lines. One goes from P to Q across the front of the sphere; the other goes around the back of the sphere, circling around to Q. Both are "straight" lines between P and Q.

Surely the meaning of 'straight' changes from Euclid's use to that of elliptic geometry. Apart from the foregoing departure from Euclid's first postulate and reinterpretation of his second, as well as other violations, elliptic geometry changes the *paradigms* used to "*ground*" straightness. Extension is determined in part by paradigms: "*This* is straight, *that* is not" (pointing). And speakers counted circles around spheres "*non*straight," at least in the contexts of interest here. Only with the advent of non-Euclidean geometry did the great circles on a sphere come to serve as examples of "straight lines": Before that, they would have been used precisely as foils, in order to compare "straight" lines with others.[5]

With a change in the meaning of 'straight' came a change in the meaning of 'triangle', because a triangle is formed by straight line segments. It is only because great circles are now counted as straight lines that a figure formed on the surface of a sphere with angles summing to 270° can count as a *triangle*.

Because the meaning of 'triangle' changed with the advent of non-Euclidean geometries, speakers might well have concluded after being presented with non-Euclidean geometries that geometers of the Euclidean period had been right to say "Any triangle's angles sum to 180°." They had been right despite the alleged counterexamples like the non-Euclidean "triangle" with angles summing to 270°, because the alleged counterexamples fail: They are not really "triangles" at all in the original sense of the word. A figure on a sphere or other nonflat surface can be described by 'triangle' only if the meaning of 'triangle' is changed. As it happens, speakers concluded, when presented with non-Euclidean geometries and general relativity, that earlier speakers had been mistaken to assert "Any triangle's angles sum to 180°." But they might have concluded otherwise.

These geometrical examples fit a pattern familiar from earlier chapters. Sentences using 'triangle' have been rejected after theory change and meaning change. Quineans will insist that in cases like this, meaning change and theory change are indistinguishable. If that is right, then my own arguments from previous chapters would also seem to undermine the distinction between meaning change and theory change. My examples are obviously similar. I have argued that formerly accepted sentences using 'dinosaur', 'rodent', 'fish', and other such words have also been rejected after theory change and meaning change.

Certainly it would have seemed obvious to earlier speakers gazing upon a reconstructed *Apatosaurus* (*Brontosaurus*) specimen that 'Birds are dinosaurs' is false. Now we know that birds are descended from dinosaurs. Thus, biologists of the cladistic school announce that birds are dinosaurs. This conclusion does reflect theory change. The birds have been found to have descended from dinosaurs. Because there has been theory change, the conclusion that birds are dinosaurs does not seem just groundless.

Even so, speakers polished the meaning of 'dinosaur' in concluding that the birds are dinosaurs: They could have continued to insist after the empirical disruption that "Birds are not dinosaurs," and been no more out of sync with earlier use. Speakers could have said, for example, that because birds rapidly evolved after taking to the air, they are not members of the dinosaur kind, which is more natural without them. That is in essence the argument evolutionary taxonomists have, in fact, offered, and it does not seem simply *wrong*. Or, speakers could have simply dismissed 'dinosaur' from scientific duty, announcing that because scientific duty (for cladists) would demand that birds are dinosaurs and because this choice seems odd, 'dinosaur' names an artificial group. Again, lots of terms have gone this way: see Chapter 3 for examples.

154

There has been meaning change as well as theory change. Quineans say that meaning change cannot be distinguished from theory change. My examples appear to add more testimony to the Quinean case against any distinction.

Quineans insist that even statements that seem true or false by *definition* may turn out otherwise after conceptual disruption. Revision in geometry nicely illustrates this point. So do various statements about natural kinds, as I have shown. The truth of 'New species cannot evolve from ancestors' seemed to earlier speakers to be guaranteed true by the very definition of 'species'. Recall Watson's claim that "The very definition of the term 'species' as usually given, involves an assumption of non-transition," or Hopkins's claim that "Every natural species must by definition have had a separate and independent origin."

Yet even what appeared to be true by definition has been rejected. Scientists have followed Darwin in rejecting 'New species cannot evolve from ancestors'. There has evidently been some meaning change, as Darwin concedes when he agrees that "in the ordinary acceptation of the word" two lineages taken for separate species would be counted one if a common ancestry could be proven (1975, p. 96). But a change in theory was also responsible for the revision. For Quineans, theory change and meaning change cannot be distinguished in such cases of revision. Moreover, any statement can be revised by this sort of theorymeaning change.

### I.2. *Theory Change, Then Meaning Change*

Quineans are right, then, to doubt in cases like the foregoing that "the dichotomy 'meaning change *or* theory change' is tenable" (Putnam 1975e, pp. 255–6). There has been both meaning change *and* theory change. But it is another step entirely to say that meaning change and theory change are *indistinguishable*. Quineans' examples have prompted many to make that further step of saying that theory change and meaning change are indistinguishable, but those examples hardly support such a strong claim.

How can one distinguish theory change from meaning change in scientists' conclusion that "New species evolve from ancestors"? One can distinguish by saying that the theory change came first. Scientists came to see that the earlier use of 'species' was unfortunate in that it did not hook up to the world in a helpful way. After that came the meaning change. Scientists altered the meaning of 'species' to make it hook up in a helpful way. Elaboration will help clarify the steps.

I.2.a. THE THEORY CHANGE. Earlier speakers seemed to use 'species' to stand for groups that were supposed to meet more than one condition. They were supposed to meet both (a) the condition of being like those entities that naturalists *called* "distinct species," for example the polar bear, the rainbow trout, and the radish and (b) the condition of including all blood relatives of any member. Each of these conditions was supposed to be necessary and sufficient for the application of the term, because scientists supposed that whatever satisfies one condition has to satisfy the other. They were wrong. No group satisfies both conditions.[6]

What, then, should be said of the polar bear, the rainbow trout, and the radish: Are they species? I propose that the theory of evolution showed that 'species' was *vague*. The vagueness in 'species' and other theoretical terms is multidimensional. Some terms are saliently vague along a single dimension. Consider 'tall'. A woman of seven feet is surely tall. A woman of four feet is not tall. But there is a vague range in which it is not clear whether a woman in that range is tall. A woman who is 5'8" is a borderline case: It is not clear whether she is a tall woman.

'Tall' is saliently vague because of gradual variation along one dimension. Other terms are saliently vague on account of variation along many dimensions. 'Religion' is an example. Clear cases of religions have several features in common, including belief in a deity or deities, prayer and other forms of interaction with the supernatural, a view of the significance or "meaning" of the world as a whole, a social organization built around these beliefs and practices, and so on. Roman Catholicism has all of these features. It surely seems to qualify as a bona fide religion. A group of people waiting at a bus stop lacks all of these criteria and fails to qualify. But it is not clear just what combination of features is necessary or sufficient for qualification. Hinayana Buddhism, for example, rejects belief in supernatural beings, as does Communism; yet both have been called religions. Whether they are religions seems unclear. 'Religion' is a word with multidimensional vagueness. Other such terms include 'big' and 'nice'. Different dimensions count in determining whether something is big: height and volume. 'Nice' is similar, because people can score well in some departments but not others.[7]

'Species' seems to have had vague application because of this type of multidimensional vagueness. Two dimensions, in particular, came into play. Although the polar bear, the rainbow trout, and the radish each met *one* condition associated by nineteenth-century naturalists with the term 'species', that of being (a) relevantly like groups naturalists called "distinct species," they failed to meet another condition, that of being (b) inclusive of all blood relatives of any member. This made for vagueness. The lineages to which naturalists had pointed, saying, "Here is surely a distinct species," turned out to be vague cases: They met with some but not all conditions associated with the term 'species'.

The discovery of borderline cases marked *theory* change, not meaning change. People became aware when they accepted evolution that 'species' was vague. This vagueness caused it to be unclear whether speakers should drop the claim that polar bears and rainbow trout are distinct "species," reporting "Polar bears and rainbow trout are blood relatives, so they are not distinct species," or whether speakers should drop the claim that species include all blood relatives of any member, reporting "New species evolve from ancestors." The statement 'New species evolve from ancestors' failed to be straightforwardly true. Instead, the sentence was indeterminate, or true to a degree, or partially true, or true on some but not all precisifications, or whatever such vague sentences are.[8]

I.2.b. THE MEANING CHANGE. Speakers *could* have gone ahead and lived with the vague application of 'species'. But that would have rendered the term useless. What speakers wanted to talk about were lineages like the polar bear, the rainbow trout, and the radish lineages, which inconveniently turned out to be borderline "species." So a term was needed for them. 'Species' would have sufficed, if assigned a new, slightly different meaning, or another term would have done instead, in which case 'species' could have been retired because of the connection to common descent. The scientific community chose the former alternative, following Darwin's lead. The community altered the meaning of 'species'.

Given that nothing ended up meeting both conditions, Darwin removed one condition. He dissociated from the term 'species' condition (b), that any "species" include all blood relatives of any member. He chose instead to associate the term 'species' only with condition (a), that any "species" be like those entities that naturalists *called* "distinct species," for example the polar bear and the radish. Darwin was explicit about this refinement. Early in *Natural Selection* he writes, "In the following pages I mean by species, *those collections of individuals, which have commonly been so designated by naturalists*" (1975, p. 98; my emphasis).

This is where *meaning change* enters. It is separate from theory change, and it follows theory change. On the use of 'species' that resulted from Darwin's refinement, it is perfectly appropriate to say that species can evolve. Nothing in the concept expressed by 'species' makes such a claim dubious or inappropriate anymore.[9]

In severing 'species' from condition (a), Darwin gave up what had been considered a defining principle of 'species', as Quineans would emphasize. He gave up the defining principle that any species include all blood relatives of any member. The reason Darwin felt right to surrender this "defining" principle is that in practice there was another defining principle, whether or not this was made explicit by any speakers. When the different principles were found to apply to different groups, the one condition did not simply trump its competitor and determine reference. Instead, the conflict resulted in vagueness.[10]

### I.3.  *Open Texture and Cold Change*

It may be helpful to compare the changes that attended 'triangle' or 'species' with what might be called "cold change," or linguistic change that is *not* made in response to conceptual upheaval. Such a comparison will help to show that conceptual change of the kind I have considered at length for the past four chapters does not undermine the distinction between theory change and meaning change. Consider the refinement of ordinary vagueness that speakers have known about for some time. 'Hot drink' has such vagueness. 'Hot drink' is vague because a boiling drink of 212 degrees F is hot, and a drink of 40 degrees F is not hot, but there are cases that do not clearly count or fail to count as hot – for example, a drink of 110 or 120 degrees F. Suppose that the vagueness becomes bothersome, so we decide, unprompted by conceptual upheaval, or any need to accommodate new information, to eradicate it. We could tell some story to explain the motivation: Perhaps we run a restaurant, and we are often asked how hot our hot drinks are, so we stipulate a use of 'hot drink'.

We announce, "The line between what we shall call a 'hot drink' and what we shall call a 'non-hot drink' is to be 110 degrees F." This is a pure stipulation. There is nothing special about the number 110: It is just chosen by lots from several possible numbers ranging from around 110 to around 130 so that some precise number or other can be assigned to the label 'hot drink'.

Here there appears, at least prima facie, to be no theory change at all. Although a philosopher driven by the theory that theory change and meaning change are inseparable might insist that there is theory change here, no one would be *motivated* by considering such cases to accept *in the first place* the theory that theory change and meaning change are inseparable. No one would be prompted to accept the view that theory change and meaning change are inseparable by such an example because the example seems on its face to illustrate pure meaning change. We have simply polished the meaning of 'hot drink' to eradicate vagueness. Similar polishings can be performed for 'bald', 'red', 'religion', and so on. Because 'religion' is vague with respect to whether belief in a supernatural being is required, we could *decree* that it *is* required and thus that Communism and some forms of Buddhism are simply not "religions" in our polished sense of the word.

Such refinement of vagueness that has long been recognized does not provide convincing grounds to reject the distinction between theory change and meaning change. But if that is the case, then the kind of change that has attended words like 'triangle' and 'species' cannot provide convincing grounds to reject the distinction between theory change and meaning change, either. The linguistic change is relevantly similar in the two cases.

The important difference between cold change and the change that attends terms like 'triangle' and 'species' seems to be just that in the latter case the vagueness came as a *surprise*: It was open texture. Before the common acceptance of evolution, people did not realize that 'species' had borderline application to the polar bear, the rainbow trout, the radish, and so on. The various criteria associated with 'species' were supposed to apply together. Scientists had to inform us that they did not. Vagueness had to be revealed. By contrast, scientists have not had to inform speakers of the ordinary vagueness characterizing terms like 'hot drink' or 'nice' or 'red'. We have known all along that red grades off into orange through borderline shades; scientists did not have to tell us that.

As I have said, the important difference between cold change and the change that attends terms like 'triangle' and 'species' seems to be just that in the latter type of case the vagueness has not been known about all along. But, of course, how long vagueness is recognized before it is polished out of a word should hardly matter to whether the polishing itself is a pure meaning change or a theory change too. If the one case does not provide convincing grounds for taking there to be no distinction between theory change and meaning change, the same should be said about the other case.

## II.   NECESSITY, THE CAUSAL THEORY, AND ANALYTICITY

In the previous section, (I), I argued that linguistic revision that follows conceptual upheaval does not provide convincing grounds for rejecting the distinction between theory change and meaning change. Recognizing such conceptual change, as I do and as Quineans do, does not commit one to rejecting the difference between theory change and meaning change.

In this section I argue that most philosophers are committed to *endorsing* the distinction between theory change and meaning change. This has been the case for some thirty years, though no one seems to have noticed. Most philosophers are committed to endorsing the distinction between theory change and meaning change because most philosophers accept the necessity associated with the causal theory of reference. Most philosophers accept that the statements 'Hesperus = Phosphorus' and 'Cicero = Tully' are necessarily true because they are true and because 'Hesperus', 'Phosphorus', 'Cicero', and 'Tully' are rigid designators. That some statements like 'Hesperus = Phosphorus' are necessarily true is decidedly the dominant position.

In the remaining sections of this chapter I will accept as a premise that statements like 'Hesperus = Phosphorus' are necessarily true, as most philosophers grant.[11] I will argue on the basis of this premise that there is an analytic–synthetic distinction. It follows from there being an analytic–synthetic distinction that there is a theory–meaning distinction. Before the argument, I must clarify the account of analyticity that is at issue.

## II.1. *A Tenable Dogma*

Analytic statements are to be characterized at a first approximation as statements that are necessarily true by sole virtue of word meanings and, at least on the version of analyticity that is of interest here, logical truth. They are thus marked by a certain triviality. 'Bachelors are unmarried', 'A somnambulist is a sleepwalker', and 'A vixen is a female fox' are traditional examples of analytic statements. Synthetic statements are non-analytic. 'Bachelors are permitted by law to vote', or 'Vixens typically raise six cubs at a time' are traditional examples of synthetic statements. Intuitively, these sets of examples mark a fundamental distinction. But Quine has made it popular to deny that there is any such distinction to mark. Thus, Quine: "That there is such a distinction to be drawn at all is an unempirical dogma of empiricists, a metaphysical article of faith" (1961, p. 37).[12]

The account of analyticity that I discuss here depends crucially on synonymy. Quine targets synonymy-generated analyticity in "Two Dogmas of Empiricism." There Quine distinguishes between two classes of allegedly analytic statements.

Those of the first class, which may be called *logically true*, are typified by:

(1) No unmarried man is married.
The relevant feature of this example is that it not merely is true as it stands, but remains true under any and all reinterpretations of 'man' and 'married'. If

we suppose a prior inventory of *logical* particles ... then in general a logical truth is a statement which is true and remains true under all reinterpretations of its components other than the logical particles.

But there is also a second class of analytic statements, typified by

(2) No bachelor is married.
The characteristic of such a statement is that it can be turned into a logical truth by putting synonyms for synonyms; thus (2) can be turned into (1) by putting 'unmarried man' for its synonym 'bachelor'. (Quine 1961, pp. 22–3)

Quine's target in "Two Dogmas" is the latter. "Our problem," he announces after drawing his distinction, "is analyticity; and here the major difficulty lies not in the first class of analytic statements, the logical truths, but rather in the second class, which depends on the notion of synonymy" (p. 24). Synonymy, Quine argues, is not up to the task of generating analytic truth for sentences like 'No bachelor is married' even *given* the trivial analytic truth of the corresponding sentence 'No unmarried man is married'.

I will argue that if we agree that statements like 'Hesperus = Phosphorus' and '*Brontosaurus* = *Apatosaurus*' are necessarily true, then we are committed to recognizing synonymy. Indeed, we are committed to recognizing *statements that are necessarily true by sole virtue of synonymy* plus logical truth. Hence, we are committed to recognizing analytic statements.

Analytic statements are *unrevisable* in an important respect: They cannot under any metaphysically possible circumstances be found to be false. Because an analytic statement cannot be false, anyone who takes it to be false or in need of correction is simply mistaken.[13]

So an analytic statement is unrevisable in one substantial respect: There is no metaphysical possibility that it be found to be false or in need of correction. Still, no analytic statement or sentence is unrevisable in just any respect. For starters, an analytic sentence may come to mean something else, in which case speakers could be right to reject it. *Any* sentence, including one that states a necessary truth, can state a falsehood if it loses its original meaning. Even 'Hesperus = Phosphorus' will be false if it comes to express a falsehood – for example, if it comes to state that W. V. Quine was born before Charles Darwin was. The same applies to analytic sentences, of course, which are just one species of necessarily true sentence. All sentences are revisable in the respect that they can be false on some interpretations. It will be helpful to have a term for a sentence *with an assigned meaning*. 'Statement' will do. On this use, an analytic *statement* cannot change its meaning and become false because a statement, unlike a sentence, cannot change its meaning.

Second, and more critically, an analytic statement need not be unrevisable in the respect that speakers know with skeptic-proof certainty or infallibility that it is true and hence not to be given up. We can never rule out the possibility of error. Even truths of mathematics and logic are subject to skeptical worries. It is possible, for all an agent may be able to satisfy herself before the skeptic, that even '7 + 5 = 12' is false. The agent may be unable to conceive how it *could* be false, but that does not preclude the possibility. The agent's inability may be explained by natural cognitive limitations imposed by evolution. Or the inability may be explained by the absence of any alternative arithmetic analogous to the non-Euclidean geometries that affirm so many possibilities inconceivable to earlier generations. Or the agent's inability to see how '7 + 5 = 12' could be false may be explained by the clever arts of an evil demon or a mad scientist intent on deceiving the agent. The skeptical possibilities are endless. Even our most certain beliefs are subject to skeptical doubts. So there is no reason to suppose that *analytic* belief is somehow uniquely protected against such skeptical possibilities.

Nor need an analytic statement be such that giving it up would always be *irrational*. Giving it up would always amount to giving up a true claim, but that is a different matter. Testimony, for example, might prompt a rational agent to give up the belief that bachelors are unmarried, even if 'Bachelors are unmarried' is analytic. Elliott Sober's Wise One story shows this (Sober 1984, p. 66: Sober credits the case to Philip Kitcher):

> Suppose we know someone who is an extremely insightful and trustworthy authority. We also know this person to be very honest. This Wise One says to us one day, "philosophers are always saying that all bachelors are unmarried. But if you look very carefully at what these concepts mean, you'll see that this doesn't have to be true. And, in fact, there are some bachelors who are not unmarried." Now I suggest that it would be *pigheaded* to simply dismiss the remarks of the Wise One out of hand. People have made mistakes in analyzing concepts before, and if the Wise One is so smart and honest, we ought to take him seriously in the present case.

Sober is right. It would be rational to back off from a belief that bachelors are unmarried in the face of such testimony from the Wise One. We are fallible creatures and would do well in situations like this to remember that. But this should not embarrass proponents of the analyticity at issue. Even if we can never rule out the possibility that we have made an error in thinking that a statement is analytic – because, say, our judgements about synonymy could

162

be mistaken for all that we can be sure – it may still be that some truths are in fact analytic.[14]

Finally, although analytic truths are unrevisable in the respect that they are necessarily true, they are not necessarily true independently of any facts about the world. An analytic statement is true by virtue not only of synonymy but also logical truth. And logical truth seems to be true in virtue of extralinguistic facts about the world. If this is right, then analytic truths also depend for truth upon facts about the world as well as synonymy.

For similar reasons, analyticity is not to be associated with truth by convention. Conventions seem to have something to do with synonymy, but the truth of an analytic sentence depends also on logical truth. Hence, the truth of such a sentence is no matter of convention if logical truth is not, as it seems not to be.[15]

The version of analyticity at issue is therefore modest by comparison with some others. Nevertheless, it commands interest. One reason that it commands interest is that, as I have observed, statements that are marked by the analyticity at issue are necessarily true and are therefore unrevisable in the respect that there is no metaphysical possibility for anyone to find them to be false and make corrections. If there are analytic sentences, they are attended by significant unrevisability.

A related reason that the analyticity at issue commands interest is that if it exists it calls for a distinction between theory change and meaning change. The two changes must be distinct, and it must be possible to have one without the other, given analyticity. So a defense of this form of analyticity is a defense of that distinction. That this form of analyticity calls for a distinction between meaning change and theory change is not controversial: Quineans reject such analyticity precisely because it does call for a distinction between meaning change and theory change. But why does this form of analyticity call for the distinction? The reason is that if a sentence is analytic, then it is unrevisable in the respect that one could never by a change of theory *correctly* come to reject what it says *on its current meaning*: After all, what it says on its current meaning is not only true but *necessarily* true, or true in all possible worlds, and hence there is no possibility of falsity. But, of course, one could always *make* a truly analytic sentence false by giving it a different meaning: One could cause the sentence to express, for example, *that Quine was born before Darwin*, in which case the sentence would express something false. After changing the sentence's meaning in this way, one could come to reject the sentence, correctly, on its new meaning, given which the sentence is false. Yet if someone were to change the meaning of the analytic sentence and then were to reject the sentence on its new, false interpretation in this way, this would be a *mere* change of *meaning*. It could not be a correct change of *theory* according to which the analytic sentence is false on its *original* meaning. That is because, again, no correct change of theory according to which the sentence on its original meaning is false would be metaphysically *possible*: Being analytic, the sentence on its original meaning would be *necessarily true* and so could not in any metaphysically possible situation be false or correctly declared false. So, given the analyticity in question, any correct judgment that an analytic sentence is false would be a result of a simple meaning change that does not bring about any change in theory concerning what the sentence says on its original meaning.[16] Hence, given the analyticity in question, there must be a difference between a meaning change and a theory change.

If there are analytic statements, meaning change must be distinguished from theory change. That is a deeply interesting mark of the analyticity in question. The analyticity in question is also of interest for work it promises to do in epistemology. Knowledge of necessity has puzzled philosophers since antiquity. Some such knowledge seems to be about analytic truth – for example, the apparent knowledge that 'All somnambulists are sleepwalkers' is necessarily true. If indeed this *is* knowledge, how is it possible for us to have it? Certainly no survey of somnambulists would yield any such knowledge. If the statement at issue is analytic, then that provides progress in answering the question: The knowledge is achieved by a grasp of synonymy and the necessity of logical truth.[17]

## II.2.   *A Commitment to Synonymy and Analyticity*

That is the analyticity at issue. In what follows I address its relationship, both real and perceived, to the necessity associated with the causal theory of reference.

II.2.a.   A PERCEIVED NONCOMMITMENT TO ANALYTICITY. That philosophers who accept Kripke's arguments for the necessity of 'Hesperus = Phosphorus' must accept the analytic–synthetic distinction, as I will argue that they must, is hardly widely recognized. On the contrary, one of Kripke's achievements is supposed to be precisely that he shows, using the causal theory of reference, that we can have necessity *without* analyticity. When he attacked analyticity, Quine did not distinguish analyticity from necessity and apriority. Therefore, Quine appeared to cast into doubt all three notions. Much of the immense interest in Kripke's arguments for necessity arises on account of Kripke's having *distinguished* these notions.[18] The necessity associated with the causal theory of reference is *not* analytic, proponents insist. It is not even a priori: Empirical inquiry is needed to determine that 'Hesperus = Phosphorus' is necessarily true. Thus, the causal theory of reference *appears* to allow us to help ourselves to necessity even while acknowledging the soundness of Quinean objections to analyticity.

The popularity of the causal theory of reference has contributed substantially to the general *rejection* of analyticity, in part for the reasons expressed above: In view of the necessity associated with the causal theory, analyticity appears to many to be dispensable, because we can have necessity without it. Besides rendering analyticity dispensable, the causal theory appears to many to provide more direct reasons for abandoning analyticity. The range of causally grounded words may leave no place for terms with old-fashioned analytic definitions. Kornblith says for this reason that the theory of reference developed by Kripke, Putnam, and others provides "good reasons for doubting the tenability of the analytic/synthetic distinction" (1980, p. 110). And Sidelle observes that the causal theory of reference is one of the "basic sources for the scorn, suspicion, and smirks with which talk about analyticity is commonly met in much of the contemporary philosophical community" (1989, p. 136).

As Sidelle suggests, philosophers remain highly suspicious of analyticity, but they generally acknowledge necessity, and in particular a posteriori necessity.[19] The causal theory of reference has encouraged this combination of commitments. The combination is confused.

II.2.b.  WHY THERE IS A COMMITMENT TO ANALYTICITY. The truth is that the necessity associated with the causal theory cannot hold unless analyticity is tenable. If we recognize that 'Hesperus = Phosphorus', 'Cicero = Tully', or '*Brontosaurus = Apatosaurus*' is necessarily true, then we must recognize that other statements are necessarily true because they contain two synonyms. Consider: According to the new theory, a statement like '*Brontosaurus = Apatosaurus*' is necessarily true (if true at all) because '*Brontosaurus*' and '*Apatosaurus*' both designate a particular dinosaur kind rigidly, or in all possible worlds. Because both terms designate just that dinosaur in any possible world, the statement '*Brontosaurus = Apatosaurus*' is true in each possible world. It is necessarily true. People might mistakenly reject it, as the paleontologist O. C. Marsh did (see Chapter 2), but they cannot find out that it is false. This is just Kripke's familiar story from *Naming and Necessity*.[20]

A *synonym* for '*Brontosaurus*' would generate necessary truth in the same way, if synonymy is a notion that we can accept. Quine finds trouble with the notion. Yet it seems very difficult to allow that '*Brontosaurus*' and '*Apatosaurus*' refer to the very same thing in each possible world while *dis*allowing synonymy. Let us assume at the outset that '*Brontosaurus*' and '*Apatosaurus*' are *not* synonyms (though in fact biological systematists refer to them as 'synonyms'). Let us say that in order for '*Brontosaurus*' and '*Apatosaurus*' to qualify as synonyms, Marsh would have had to have coined them with that in mind, saying something like "Let '*Apatosaurus*' and '*Brontosaurus*' be synonymous terms for such-and-such," or "Let '*Brontosaurus*' mean *Apatosaurus*." Marsh did nothing like this because he did not intend for his two terms to share reference conditions.

Even if synonymy is attended by requirements like the foregoing, synonymy must be possible given that there can be more than one rigid designator for an object. '*Brontosaurus*' refers to *Apatosaurus* in all possible worlds because Marsh decreed, in effect, "Let '*Brontosaurus*' name the genus of dinosaur that left these fossil remains," while pointing to fossils left by a specimen of the genus *Apatosaurus*. If that is possible, then Marsh could certainly have coined another term along *with* '*Brontosaurus*' as a *synonym*. He could, for example, have coined '*Thunder-Lizardosaurus*' and '*Brontosaurus*' as synonyms at the same time, by decreeing, "Let *both* terms '*Brontosaurus*' and '*Thunder-Lizardosaurus*' name the genus of dinosaur that left these fossil remains." Such dubbing is clearly permissible, given the familiar causal account of how terms like '*Brontosaurus*' come to refer rigidly. It is clearly permissible because according to the relevant account of reference, a dubber can coin as many different terms for something as she likes, using the same physical object to ground the baptism: In this way she can assure that several terms rigidly designate the same thing. Indeed, speakers used the very same physical object to coin 'Hesperus' and 'Phosphorus'; that is how these two different terms came to be rigid designators for the same thing. Speakers did not *know* they had coined two different names for the same object by pointing to that same object twice, but clearly such ignorance is not *needed* for speakers to coin two different names for the same object by pointing to that same object twice.

So given the familiar causal account of how terms like '*Brontosaurus*' come to refer rigidly, it seems clear that Marsh could have coined '*Thunder-Lizardosaurus*' as a synonym for '*Brontosaurus*'. This synonymy would generate analyticity: '*Brontosaurus* = *Thunder-Lizardosaurus*' would be necessarily true given the necessity of the logical truth '*Brontosaurus* = Brontosaurus*', which proponents of the necessity associated with the causal theory of reference certainly must grant, and given the relevant synonymy.[21] In the same way, statements like 'Every *Brontosaurus* specimen is a *Thunder-Lizardosaurus* specimen', 'No *Brontosaurus* specimen could fail to be a *Thunder-Lizardosaurus* specimen', and so on would be analytic. Because the synonymy needed to generate analyticity is possible, analyticity must be tenable.

A more familiar way of coining synonyms than the process described here would be to state explicitly something like "Let '*Thunder-Lizardosaurus*' be another name for *Brontosaurus*." This method of coining synonyms is also clearly possible given the necessity associated with the causal theory of reference. For if Marsh could get synonyms by decreeing, "Let both '*Brontosaurus*' and '*Thunder-Lizardosaurus*' name the genus of dinosaur that left these fossil remains," then there could hardly be any barrier to his getting the desired synonymy by saying, instead, "Let '*Brontosaurus*' name the genus of dinosaur that left these fossil remains, and let '*Thunder-Lizardosaurus*' be another name for that genus," or "Let '*Brontosaurus*' name the genus of dinosaur that left these fossil remains, and let '*Thunder-Lizardosaurus*' be another name for *Brontosaurus*." Clearly *any* of these options works if another does.[22] If you accept the necessity associated with the causal theory of reference, you simply must admit that synonymy is a perfectly legitimate notion.

II.2.c. A COMMITMENT TO LASTING SYNONYMY. A commitment to the necessity associated with the causal theory of reference is a commitment to synonymy. Still, for all I have said so far, fans of the necessity associated with the causal theory of reference could object to *lasting* synonymy. Quineans sometimes acknowledge fleeting synonymy without acknowledging long-term synonymy.[23] Is any relief to be found here for Quineans? It is highly doubtful. Proponents of the necessity associated with the causal theory of reference seem not to be entitled to put Quinean restrictions on synonymy's duration.

Proponents of the necessity associated with the causal theory of reference maintain that 'Hesperus = Phosphorus' and '*Brontosaurus* = *Apatosaurus*' have remained necessarily true over time. These sentences have not quickly ceased to express any necessary claim as a result of any quick fading away of the relevant rigid designation. The reference of '*Brontosaurus*' and '*Apatosaurus*' to the same genus has obtained since 1879, to say nothing of the reference of 'Hesperus' and 'Phosphorus'. So it is hard to see why a synonym could not designate *Apatosaurus* rigidly over a long period of time as well. To reject the longevity of synonymy while accepting the longevity of rigid designation for terms coined as the causal theory of reference describes would force one to accept the bizarre consequence that so long as '*Brontosaurus*' and '*Apatosaurus*' are coined *separately*, as in the separate baptisms "Let '*Brontosaurus*' name the genus of dinosaur that left these fossil remains" (the speaker points), and "Let '*Apatosaurus*' name the genus of dinosaur that left these fossil remains" (the speaker points again), then each term can enjoy a long tenure as a rigid designator for *Apatosaurus*. But if the terms are coined *together*, as in the baptism "Let '*Brontosaurus*' and '*Apatosaurus*' both name the genus of dinosaur that left these fossil remains" (the speaker points once) then, as the terms would be synonyms, at least one of those terms could not enjoy a long tenure as a rigid designator for *Apatosaurus*. This is surely implausible. It would seem that a commitment to the necessity associated with the causal theory of reference is a commitment to full, lasting synonymy.

### III. ANIMADVERSIONS TO ANALYTICITY CROSS NECESSITY ALSO

If the arguments from the preceding section stand up to scrutiny, then generally acknowledged, plausible, modest instances of necessity associated with the causal theory of reference compel us to accept analyticity. If we accept these instances of necessity, then we must relinquish the familiar use of two slogans against analyticity: 'There is no difference between a change of meaning and a change of theory' and 'There is no guarantee against revision for any statement'. Proponents of the necessity associated with the causal theory of reference are committed to analyticity, so they are committed to a distinction between theory change and meaning change and to the unrevisability, suitably understood, of some statements. There is more. Proponents of the necessity associated with the causal theory of reference have further, independent reason to relinquish the familiar use of the foregoing slogans against analyticity. Even if, contrary to the arguments of the previous sections, proponents of the necessity associated with the causal theory of reference were not *committed* to analyticity, they would still have to deny that the slogans have any *force* against analyticity. So I will argue in the present section.

   Consider the slogan 'There is no difference between a change of meaning and a change of theory'. Proponents of the necessity associated with the causal theory of reference cannot agree. If meaning change and theory change were really indistinguishable, then it would not have been possible to discover, as Kripke would insist that we can, the necessity of '*Brontosaurus* = *Apatosaurus*' or 'Hesperus = Phosphorus'. After all, discovery, or what Kripke means by 'discovery', is a mere change of theory, not a change of meaning. Kripke insists that there is no meaning change when communities come to affirm the truth of a posteriori, necessary statements like the foregoing (Kripke 1980, p. 138. Putnam 1975e, pp. 224–5 follows Kripke here, though he apparently fails to see where this leads, as I will indicate a couple of paragraphs further into this section). If there were no difference between meaning change and theory change, then discovering that '*Brontosaurus = Apatosaurus*' is true could not be distinguished from changing the meaning of '*Brontosaurus = Apatosaurus*', and that would frustrate the Kripkean claim that scientists have discovered necessary statements like this to be true, rather than having changed the meanings of theoretical terms in the sentence.

So if you take '*Brontosaurus = Apatosaurus*' to have been discovered to be true, you must reject the slogan 'There is no difference between a change of meaning and a change of theory'. Indeed, you must reject the slogan if you so much as take '*Brontosaurus = Apatosaurus*' or 'Hesperus = Phosphorus' to be *necessarily true*, *regardless* of whether you take it to have been *discovered* to be so. No right-headed change of theory can result in the *correct rejection* of a necessarily true statement. Suppose '*Brontosaurus = Apatosaurus*' is necessarily true (though a posteriori investigation might be needed to reveal its necessary truth). In that case, there is no metaphysical possibility for a right-headed change of theory to result in the correct rejection of that statement. Only a simple change of *meaning* could make it metaphysically possible to correctly reject '*Brontosaurus = Apatosaurus*'. This change of meaning resulting in the correct rejection of '*Brontosaurus = Apatosaurus*' would have to be *distinguishable* from a change of theory resulting in the correct rejection of the original statement '*Brontosaurus = Apatosaurus*'. That is because a change of *theory* resulting in the correct rejection of the original statement would be metaphysically *impossible*, given the original statement's necessary truth. A change of *meaning* resulting in the correct rejection of the *sentence* would, on the other hand, be metaphysically possible. Everyone acknowledges that a meaning change could result in the use of '*Brontosaurus = Apatosaurus*' to express something false, such as, perhaps, that Quine was born before Darwin. In that case one could correctly reject the sentence, but only on account of a simple change of meaning; there would be no correct change of theory about the original, *necessarily true* statement.

Proponents of the necessity associated with the causal theory of reference are committed to distinguishing between meaning change and theory change. This commitment is largely unrecognized. The dearth of explicit discussions about the matter suggests that the question of whether causal theorists are committed to the distinction has hardly so much as been entertained by philosophers, despite the massive amount of discussion that has taken place concerning both the causal theory and the theory–meaning distinction. Champions of the necessity associated with the causal theory of reference have entertained the question, but they have not done much to call the issue to the attention of the larger community in a helpful way. When pressed in discussion, Kripke has affirmed briefly that he recognizes a distinction between meaning change and theory change ("Second General Discussion Session" 1974, pp. 513–14), but he generally avoids explicit discussion of the matter. Putnam, unlike Kripke, has discussed the matter explicitly at length. Unfortunately, Putnam's discussion is confused: Even in "The Meaning of 'Meaning'," where he argues that sentences like 'Water is H$_2$O' are necessarily true, Putnam suggests that the distinction between meaning change and theory change is spurious.[24] Putnam is not entitled to say this. There is no metaphysical possibility that a right-headed change of theory could result in the *correct rejection* of a necessarily true statement. So if 'Water is H$_2$O' is necessarily true, then there is no metaphysical possibility that a right-headed change of theory could result in the correct rejection of that statement. Only a simple change of *meaning* could make it metaphysically possible to correctly reject 'Water is H$_2$O'. This change of meaning resulting in the correct rejection of 'Water is H$_2$O' could not be, and therefore would have to be distinguishable from, a change of theory resulting in the correct rejection of the original statement. Again, a change of theory resulting in the correct rejection of the *original statement* would be metaphysically *impossible*, given the statement's necessary truth: A necessarily true statement cannot be false, so it is not subject to correct rejection.

The necessity associated with the causal theory of reference calls for a distinction between meaning change and theory change. Further, it would seem that other varieties of necessity similarly call for that distinction. In particular, even traditional, a priori necessity calls for a distinction between meaning change and theory change. If it is indeed necessarily true that $7 + 5 = 12$, then this is true in all possible worlds: It could not possibly be false or correctly rejected under any possible circumstances. Hence, if it is necessarily true that $7 + 5 = 12$, then there is no metaphysical possibility that we might – after discovering an alternative arithmetic, say – *correctly* reject what '$7 + 5 = 12$' says on its original meaning. We could only assign a new meaning to '$7 + 5 = 12$' to make the sentence say something false, and *then* correctly reject the sentence on the new meaning. This change of meaning resulting in the correct rejection of '$7 + 5 = 12$' could not be, and therefore would have to be distinguishable from, a change of theory resulting in the correct rejection of the original statement. As I have emphasized, a change of theory resulting in the correct rejection of the *original statement* would be *impossible*, given the statement's necessary truth: A necessarily true statement cannot be false, so it is not subject to correct rejection.

So much for the slogan 'There is no difference between a change of meaning and a change of theory'. How does the slogan 'There is no guarantee against revision for any statement' fare as an attack on analyticity? This slogan has a plausible air about it, no doubt because it is *true* on *some* interpretations, or given some uses for 'revisable'. Whether the slogan is true on any interpretation that undermines analyticity is another matter. What is meant by saying that any statement is revisable, or that any statement can be legitimately rejected? If this is an *epistemic* claim, a mere admission to the skeptical possibility that any statement, however certain it seems, could turn out to be false for all that we can be sure before the skeptic, then it is true, but of no consequence to analyticity. It is true because even when a statement seems self-evident, error can never be ruled out. Frege probably took the axioms of his naïve set theory to enjoy maximal certainty, but he learned, to his sorrow, that a contradiction in them had escaped his notice. Ancient Greek geometers found that many of their foremost theorems collapsed after the Pythagoreans' discovery of irrationals. Sometimes what seems certain is spurious.

Still, there is no worry for analyticity here. The skeptical possibility that we could be subject to error in *thinking* that a statement is metaphysically necessary does not preclude the statement's in fact *being* metaphysically necessary, nor does it preclude the statement's in fact being necessary *by sole virtue of synonymy and logical truth*. Hence, the epistemic possibility that we are wrong in affirming a statement does not preclude its being analytic.

Suppose, on the other hand, the claim that any statement can be revised is to be construed as a claim about *metaphysical* possibility rather than *epistemic* possibility. In that case the claim is that any statement is such that there are metaphysically possible circumstances in which it is false, or in which we would not be wrong, nor merely using the sentence with some nonconventional meaning to conclude that it is false. This revisability would indeed bring dire consequences for analyticity if it attended every statement, because if every statement were revisable in this respect, then every statement would be such that there are metaphysically possible circumstances in which it is false; there could be no *necessarily true* statements, which are statements that are true in all metaphysically possible circumstances or worlds. If the claim that any statement can be revised is to be construed as a claim about *metaphysical* possibility, it follows from the claim that there could be no metaphysically necessary statements and therefore that there could be no analytic statements.

But if the thesis of revisability amounts to this strong metaphysical thesis, it is hardly the irresistible doctrine it appeared at first to be. On the contrary, anyone embracing metaphysical necessity of any sort at all is committed to rejecting it. So proponents of the metaphysical necessity associated with the causal theory of reference in particular are committed to rejecting it. For again, a metaphysically necessarily true statement is one that is true in all metaphysically possible worlds: There are thus no possible worlds or circumstances in which it is not true. And thus there are no possible circumstances in which we could reject it without error. Hence, if there is *any* metaphysical necessity, then there must be some truths not subject to the metaphysical possibility of refutation.[25]

The usual slogans brought against analyticity should not, then, have ever deterred proponents of the necessity associated with the causal theory of reference. Proponents of that theory cannot accept that meaning change is inseparable from theory change. Nor can they accept that there is a metaphysical possibility that any statement is revisable. Both claims are toxic to analyticity. But both claims are toxic to necessity also, and proponents of the necessity associated with the causal theory of reference clearly do not want to give up necessity. Not many philosophers do want to give up necessity.

Those who recognize necessity *can* accept the skeptical arguments for human fallibility. So they can accept the *epistemic* possibility that one could revise any statement and be right in so doing. But the epistemic claim of revisability is compatible with analyticity. Analytic truth is no answer to skeptics who call into question whether we know or are certain that we know any necessary truths; analytic truth is just a species of necessary truth.

## IV.   CONCLUSIONS FOR THE CAUSAL THEORY OF REFERENCE

The previous chapter and the present one both address traditional problems widely thought to be solved or finessed by the causal theory of reference. In both cases, the traditional problem is not really rendered any less troublesome.

The previous chapter addresses the problem of incommensurability. Many have supposed that incommensurability is blocked if reference is determined causally. That is because a causal account of reference is supposed to allow theoretical terms to keep their original meanings and referents through theory change. Unfortunately, however, even if reference is determined causally rather than by descriptions, reference changes with theory change. Therefore the causal theory of reference does not dispel the threat of incommensurability.

Nor does the causal theory of reference, along with the necessity associated with it, rescue necessity from Quinean problems afflicting analyticity, as the present chapter shows. Metaphysical necessity has enjoyed great popularity since the articulation of the causal theory of reference. Analytic necessity has, on the other hand, become less popular in the light of the causal theory. Quinean criticisms continue to make analyticity unpopular. Many seem to suppose that the causal theory shows how to resist analyticity and its unwanted problems while still honoring necessity: namely, the necessity associated with the causal theory. This is an illusion. If the a posteriori necessity associated with the causal theory of reference is tenable, so is analyticity. Proponents of the necessity associated with the causal theory cannot accept Quine's attack on analyticity.

It appears, then, that the causal theory of reference and the necessity associated with it has not altered the philosophical landscape nearly as much as philosophers have supposed. With or without the causal theory, conceptual upheaval can create problems for reference stability and for allegedly necessary truth. These problems are not circumvented by the causal theory of reference. They remain to be addressed. I have addressed them.

# Notes

1. Although philosophers often call all taxa "species," higher taxa like the felines or the mammals are not really species. They are more inclusive than species: Thus, both the lion species and the tiger species belong to the felines. The felines as a whole are a family, rather than a species. Species are taxa of just one rank, the species rank, but there are other ranks, including the family rank and the class rank. The term to use to refer to both species and higher taxa is not 'species' but 'taxa'.

2. The question here is not, of course, whether the empirical facts that scientists have discovered about what we now call "whales" and "fish" confirm that the sentence 'Whales are mammals, not fish' is true on its *current* interpretation. *That* question seems relatively uninteresting, because it is so easily answered in the affirmative. But philosophers who discuss essence discovery have been concerned, as I have indicated, with a more interesting question: whether scientists have found that 'Whales are mammals, not fish' is true on its *original* use, or the use that the sentence had centuries ago, when competent speakers would commonly have denied that the sentence is true.

3. The philosophy of mind, for example, may be an especially fertile area for further research in this regard: See, for example, Dennett (1994).

4. Thus, I accept a version of externalism about meaning. What a species term means has to do with the origins and properties of the type specimen. The dubber may know little about these matters.

   For a different perspective on internalism and on its relationship to some of my own research (in Chapter 4, published earlier in LaPorte 1996), see Segal (2000). Segal offers an intelligent and thought-provoking defense of internalism.

5. For similar examples, see Putnam (1975d, p. 200, and 1975f., p. 274) and Kripke (1980, p. 131). Kripke (1980, pp. 55ff.) discusses the technical notion of reference fixing.

6. Devitt and Sterelny (1999) improve upon the work of earlier proponents of the causal theory by emphasizing and better articulating the need for descriptive information. They call their preferred account a "causal-descriptive theory of reference." Devitt and Sterelny offer a nice discussion of the qua problem, which was apparently named by Sterelny (1983).

1. WHAT IS A NATURAL KIND, AND DO BIOLOGICAL TAXA QUALIFY?

1.   The dispute is not about whether there are abstract entities at all, it should be clear. No one is proposing that species cannot be kinds because there are no abstract entities. All parties agree that there are kinds; the dispute concerns only whether biological species are kinds. Moreover, some who take them to be kinds, sets, and so on wish to reduce kind-talk or set-talk, which is apparently about abstract entities, to talk about something that is less ontologically toxic. Kitcher, for example, who is an outspoken opponent of the thesis that species are individuals, takes species to be sets but would rewrite set theory in order to avoid any ontological commitment to abstract objects (1987, p. 191n.). I will set aside the issue of whether a reduction is possible and assume, for simplicity, that there are abstract entities for kind-talk to be about.

2.   Is the statement 'Lead has been generated from lighter matter' literally, or strictly speaking, true? That depends on whether what is *communicated* by the statement is also literally what it expresses. There is no need for me to commit to an answer to that: Either the statement is literally true because all that it literally asserts is that *members* of a certain kind have been generated; or else what the statement is used to communicate, namely that *members* of a kind have been generated, is true even though the statement itself, which asserts that an abstract kind itself has been generated, is not literally true. Either way there is no commitment on the part of speakers who affirm the statement to the evolution of abstract objects.

3.   See also Kitcher (1984, p. 314). Although Hull's view that species are individuals seems to be on the decline, there is growing agreement about the historical nature of species (but cf. Kitcher 1984, pp. 314–15; Ruse 1987, pp. 235–6; Stamos 1998, pp. 465–6). Thus, Sterelny: "It is now widely recognized that the contrast between kinds and individuals is not the best way of expressing the fundamental insight" behind Ghiselin and Hull's suggestion. "We should think of Hull and Ghiselin as having shown that species are historical objects," even if historical *kinds* (Sterelny 1994, p. 10).

     Note that historically restricted kinds do not seem limited to the realm of biology. Levi's 501's seem to be a kind of jeans, for example. Ghiselin (1981, p. 304) apparently bites the bullet here and refuses to recognize this sort of kind. Here and elsewhere (see also his 1980), his characterization of what things are kinds falls afoul of our ordinary concept of a kind. At times, he and Hull appear to accept this, and to claim that they are offering a new, technical vocabulary (see esp. Hull 1981). But if their use of the term 'kind' departs from conventional use, then it is not clear they are even taking issue with the thesis that taxa are kinds, in the original sense of the word 'kind', even if taxa are not "kinds" in the technical sense of Ghiselin and Hull.

4.   See note 3. Thus, many of the claims of Kripke and Putnam, who take essences to be microstructural, must be reformulated as claims about historical essences if they are to accord with standard contemporary science (for further discussion, see Chapter 3). Of course, even when scientists take genetic structure to be irrelevant to what makes the members of a species members of that species, the scientists may use genetic clues to decipher whether there is a common history, which *is* relevant to what makes the members of a species members of that species. According to some

scientists (e.g., proponents of the phylogenetic species concept; see Chapter 3), genetic structure *may*, along *with* history, be part of what makes a lineage a species. But for most biologists, a common history is at least a crucial part of what makes a given species that species, whether or not genes have anything to do with the matter. And genes have nothing *at all* to do with how *higher* taxa are delimited, at least if higher taxa are understood to be clades, as most contemporary systematists would understand them to be (see below, in the main text).

It is interesting that philosophers supporting the microstructural essence view generally indicate scant awareness of biological savants' rejection of their view, or of relevant discussions in the scientific literature. This ignorance has bewildered prominent philosophers of biology. Hull (1981, p. 290) complains that philosophers spend too much time talking about Twin Earth and not enough time listening to what biologists say about Real Earth. Ruse similarly observes that philosophers like Kripke and Putnam "generally do not bother to refer to the works of practicing taxonomists, and at times show an almost proud ignorance of the organic world" (1987, p. 227n.).

5. The ancestral group of a clade is typically understood to be a species (Hennig 1966, p. 72) that splits into two or more newly evolved species, but dubbers of clade names are also free to specify that a population or some other kind of group is to serve as the stem (de Queiroz 1992, p. 303).

6. It is strange that Ghiselin should dismiss out of hand the possibility that individuals have essential properties. Kripke (1980), for example, proposes that a *person's* origin in a particular sperm and egg is essential to him or her. The analogy to species and clades is striking. Ghiselin could, therefore, change his mind and agree that, like persons, species and clades have necessary properties *and* are individuals. Obviously, however, this would be to forfeit the argument that, because they have no defining properties, species and clades must be individuals, which is the argument that is at issue here.

7. Whether English terms like 'water' and 'salt' have enjoyed such trouble-free baptisms is another issue, one that I address in Chapter 4. One complication is that earlier speakers who baptized these words would not have had the sophisticated notion of a *chemical composition*. Here I am pointing out only that if a speaker had the notion of a chemical composition and a vial of pure $H_2O$ or NaCl, she could coin a term for this chemical kind, without any awareness of the specific formula of the substance in the vial, in the foregoing fashion.

8. Natural kinds can be found on Swoyer's (2001) list of types of properties. Other authors deny that any kinds are properties – for examples, Lowe (1997, pp. 35–6).

9. The relevant modal correspondence between *Raphanus sativus* and the kind holds as well. Because talk about *Raphanus sativus* can be interpreted as talk about species-individual *I*, it is necessary that all and only the organisms (plus organism parts, etc.) that belong to *Raphanus sativus* be the parts of *I*. It is necessary that all and only parts of *I* have the property *P* of being part of *I*. It is necessary that all and only the objects possessing *P* be the members of the species-kind *K*. Because it is necessary that the organisms belonging to *Raphanus sativus* be precisely the parts of *I*, and that these be precisely the objects with *P*, and that these be precisely the members of *K*, it is necessary that the organisms belonging to *Raphanus sativus* be precisely the members of *K*.

10. Kitcher (1987, p. 187) makes essentially the point I make above about there be-
    ing two possible interpretations. In general, Ghiselin has adamantly rejected any
    suggestion that talk about species can be interpreted as talk about kinds, classes,
    sets, and so on (see, e.g., 1981, p. 303). An anonymous referee has suggested that
    Ghiselin sometimes seems prepared to brook the suggestion that talk about species
    can be interpreted as being about kinds. Although in rare passages Ghiselin hints
    that he might be prepared to soften his position in this way (1999, p. 450), he
    surrounds such passages with hard-line passages in which he insists, on familiar
    grounds, that species cannot be interpreted as kinds (1999, pp. 449, 452, 456).
    Hull is a different story. Although a defender of the s-a-i thesis, he has suggested
    that there may be a possibility of treating species as kinds or sets. If they can be
    so treated "then we are right back where we started. Now the issue is between
    two sorts" of kinds or sets: spatio-temporally unrestricted ones, like water or gold,
    and spatio-temporally restricted ones (1987, p. 173). Species, suggests Hull, would
    have to be the latter. For all I say in this chapter, this is right, though some oppo-
    nents of the s-a-i thesis have gone on to argue that at least some species may be
    spatio-temporally unrestricted (Kitcher 1987; Ruse 1987, pp. 235–6; Stamos 1998,
    pp. 465–6).
11. For similar points, see Rosenfeld and Bhushan 2000, pp. 199–202; Wolfram 1989,
    p. 234.
12. Naturalness is world relative on an account like mine, because the named-on-a-
    Tuesday kind does have explanatory value in distant possible worlds in which laws
    apply to such kinds. The kind is natural relative to such worlds. Phil Bricker has
    also called my attention to worlds in which there are objective groupings of similar
    objects even though there are no laws or causation, so no explanations. These worlds
    contain no natural kinds, on my account, although some kinds that are natural in *our*
    world have members in such worlds. Some may prefer an analysis of natural kinds
    that is more general, so that it counts such kinds natural. On a more general account,
    natural kinds might be those with, say, any theoretically interesting property. The
    choice between these two options will not affect the classification of kinds in
    the actual world, so there is no real need to settle this matter for the sake of my
    discussion. If pressed, I would be content to specify by stipulation a use of 'natural
    kind' that has been refined a little to deal with distant worlds that leave the intuitions
    less than entirely clear. For other accounts of natural-kindhood that are related to
    mine in that they also appeal to explanation, see Bird (1998, pp. 109ff.); Kitcher
    (1984, pp. 315–16n.; 1993, pp. 169–73); Platts (1997, pp. 268ff.); Wilson (1999,
    pp. 44ff.).
13. The generalization that biological taxonomy aims to reflect the tree of life is com-
    plicated by a minority opinion that it should purge reference to descent. There are
    two well-known schools of the minority opinion, pheneticism and pattern cladism
    (for a good brief history, see Hull 2001). Pheneticism has for some time been
    out of favor. Pattern cladism also seems to be waning (see, e.g., Donoghue 2001,
    p. 756). The existence of nonhistorical schools does not undermine the account of
    naturalness that I discuss here. One reason it does not is that classification is sup-
    posed to *serve* evolutionary theorists on anyone's account. Hence, even phenetic
    or pattern cladistic taxa may be historically natural. As one pattern cladist sees
    it, "systematics provides evidence that allows inference of a scientific theory of

evolution," rather than making reference to the theory of evolution (Brower 2000, p. 143).

14. At least the barbet family is more natural ecologically if Mayr's ecological facts are right. To be sure, cladists are wont to find more in common between organisms in cladistic taxa than Mayr is: See, for example, Griffiths (1994, p. 216). Still, evolutionary taxa are likely sometimes to be more natural than cladistic taxa from an ecological perspective for reasons like the foregoing (see also note 15).

15. In the foregoing quotation Darwin is discussing the lost pedigrees of various domestic breeds, which his contemporaries acknowledge to be related by blood, in order to make a comparison with the lost pedigrees of species and higher taxa. Because Darwin is right that many important traits tend to mark historical groups, there can be practical value in a discipline as theoretical as evolutionary biology. This can be seen, for example, in some recent systematists' remark that there is "a strong economic incentive for the development of a phylogenetically based classification" of a particular pathogen that they investigate. The reason is, again, that "a natural [i.e., phylogenetic] system offers the greatest predictive value for investigating all aspects of its biology" (O'Donnell, Cigelnik, and Nirenberg 1998, p. 484). Mayr (1995, pp. 427–9) points out, however, that often early members of a clade will be more similar genetically and otherwise to closely related, coeval organisms from outside their clade than to distant descendants within their clade.

16. Ruse situates himself in a tradition that includes such figures as Whewell and Hempel. Mill belongs to the tradition as well (see, e.g., Hacking 1991, pp. 117ff.). Not everyone is enthusiastic about consilience. Ereshefsky (2001, pp. 146–7) cautions that we should not say that groups not marked by consilience are unnatural, unreal, or nonexistent. I accept this word of caution. I would carefully distinguish reality and existence from naturalness. Even unnatural kinds *exist*, and are *real*: An example would be redkind. As for whether a genealogical group is *natural* if it does not share much genetic similarity, or a common ecological niche, and so on, I would say that it is natural in one significant respect. But if a genealogical group were marked by strong genetic similarity and a shared ecological niche, and so on, it would be natural in *more* respects. I assume that each trait in question would be useful for providing explanations if it were present. To the extent that it would not be useful, its absence from a taxon does not affect that taxon's naturalness.

17. Churchland (1985, pp. 12, 13); for similar thoughts see, e.g., Collier (1996, pp. 4ff.).

18. Biological systematists almost universally follow Darwin by applying 'natural' just to historical groups. Taxa grouped on the basis of other characters, as when history is indiscernible, are called "artificial" (for some explicit examples of this use of 'natural' from diverse areas of research concerning systematics, see Bowler 1994, e.g., p. 179; Ghiselin 1987, p. 129; Gauthier, Estes, and de Queiroz 1988, p. 15; Gams 1995, e.g., p. S751). As I have already indicated, evolutionary taxonomists take into consideration more than just genealogy in their recognition of taxa, but the taxa that they recognize do have a common history.

19. The discussion of gruesome kinds helps bring to the fore an interesting question about kinds. The question is: How bound up with the idiosyncrasies of human categorization are natural kinds? It is sometimes said that natural kinds are natural independently of how humans think or group (see, e.g., Wilkerson 1993, p. 1; Ellis 2001, p. 19). On this view, a natural kind's relationship to explanation might

be worked out something like this: A kind is natural in virtue of its explanatory value from some objective, idiosyncrasy-free perspective. Others prefer to say that a kind is natural in virtue of reflecting our particular organization of nature (see, e.g., Kitcher's 1993 discussion, pp. 169–73). These two rough camps would have different perspectives concerning a group of cognizant beings that finds it more intuitive to think in terms of gruesome properties. When we report that all observed emeralds are green and all observed azurite is blue, such a group of thinkers might incline to say instead that all emeralazurite is grue, and that all azuremeralds are bleen. The one report is true if and only if the other is. But do both make use of kinds that are natural, naturalness being relative to the cognizer or group of cognizers? There is no need to commit to an answer to this question, as little hangs on the matter in the chapters that follow.

Notice that the question of whether gruekind or bleenkind is natural is different from the question whether these are members of a natural kind of *kind*, namely the gruesome-kind kind. Given the interest in gruesome kinds on the part of theorists examining induction, it seems reasonable to say that in the study of kinds, gruesome kinds comprise a natural kind of kind, and projectible kinds another.

20. A complicating factor in speaking about colors like green is that it does seem right to call the property of being green a "natural property" even in fairly strict contexts. Because greenkind is not usually called a "natural kind," because of the heterogeneity of its members with respect to properties other than color, I have to say either that the property greenness is not identical to greenkind, or else that the two are identical but that in contexts in which 'greenkind' is used the focus is on differences between members of greenkind, and in contexts in which 'the property of being green' is used the focus is on what members of greenkind have in common, namely color, so that a shift in context is responsible for the different evaluations with respect to naturalness.

21. The use of italics here may appear to be inconsistent. The convention is that a term for a species or genus is italicized, but a term for a taxon that is assigned a rank higher than the genus rank is not.

The correspondence between vernacular and scientific terms may be more or less clear. 'Tiger' and '*P. tigris*' share an extension on the *standard* use of '*P. tigris*', but some experts use '*P. tigris*' in such a way that it has a narrower extension than 'tiger', as I indicate in Chapter 3. In the event of divergence, a vernacular term may still have an extension that is as natural as the extension of a term with a closer scientific correlate. Thus, even if the extension of 'tiger' diverges from that of the species name '*P. tigris*', it is still natural: The tigers still form a homogeneous clade, in that case.

22. An ordinary reference source such as an encyclopedia will confirm this (see, e.g., "Onion"a, "Onion"b and "Garlic").

23. Occasionally, in contexts in which the subject matter is very clear, speakers might use 'lion' alone to refer to mountain lions or ant lions, or 'dog' alone to refer to hot dogs. This might show that 'lion' and 'dog' have multiple senses or, because the uses in question are clearly nonstandard, it might show only that speakers' reference sometimes diverges from semantic reference (Kripke 1977). In either case, the context here seems to behave differently from a context that sets standards for, say, the use of 'flat'. When a context determines the standard for 'flat', which

apparently has only one sense, the context seems to determine how many bumps can be properly ignored (Lewis 1983; see also 1996; Abbott 1997, pp. 316–18). When I appeal to context in the present work, it generally does not matter whether context clarifies which of multiple senses is relevant or whether it determines what can be ignored in the application of a word with one sense.

24. See also Aune (1994, pp. 57–60), who emphasizes the vagueness of our terms in a helpful discussion of precision in particular contexts.

25. Authorities who use 'lily' for the genus *Lilium* include Rockwell, Grayson, and de Graaff (1961, pp. 79–80) and Feldmaier (1970, Chapter 8). Many plants called "lilies" do not belong to *Lilium*, it is true. These include, for instance, the adobe lily, the avalanche lily, the desert lily, and the sego lily (Dupré 1993, p. 28). But all of these are simply called "false lilies" in contexts in which 'lily' is used for *Lilium*. Indeed, all are explicitly included in one authoritative text's list of "'Lilies' that are not lilies" (Rockwell, Grayson, and de Graaff 1961, pp. 331ff.). As some speakers use 'lily' for the whole family or order (see the following note), rather than restricting it to the genus *Lilium*, the foregoing examples do count as "true lilies" on these *other* more generous uses. Water lilies, on the other hand, are not called "true lilies" on any use of 'lily'.

26. Thus, Speer (1998) calls the "true lilies" the entire order Liliales, though he has agreed with my proposal that contexts of use vary (p.c.). 'Lily', vernacular though it is, is sometimes used very broadly. It might be tempting to think that even if botanists like Speer sometimes use the vernacular term 'lily' for such a wide group, lay speakers would never use the term in that way. That is not so. Lay speakers defer to science, when they have the needed information. Writers on cooking, for example, sometimes say that onions and asparagus are edible lilies (e.g., Ottobre 2001; Finn 1997). They pick up this information, no doubt, when educating themselves on the foods about which they intend to write. Note that onions, garlic, and asparagus have been removed from the Liliales recently on the basis of evidence suggesting that these groups are more distantly related than earlier taxonomists had supposed (see, again, Speer 1998). So the example is not current, but this can be ignored.

27. Thus for Dupré (1993, p. 34), scientists' evolutionary distinctions "are motivated by considerations quite unrelated to any of concern to ordinary language classification" (cf. Dupré 1999, which offers a view closer to mine on this matter). Hacking (1991, pp. 114–15; briefly discussed in van Brakel 1992, p. 250n.) similarly claims that the natural kinds recognized in common speech are not tied to the theory of evolution, as *biologists'* species are. A number of people have suggested as much to me in person at various talks. Although I provide a brief response to these claims here, I address in more detail the issue of how scientific information is relevant to vernacular use in Chapter 3, after I have provided more background information.

28. The foregoing statement was just taken from the first website on which I clicked after searching for "koala bear": www.koalaexpress.com.au/facts1.htm, p. 1 of 4. Accessed May 29, 2001.

## 2. NATURAL KINDS, RIGIDITY, AND ESSENCE

1. A few clarifications are in order. First, a rigid designator designates the same object in all possible worlds as it is used in the *actual world*, not as it is used in other

possible worlds in which the object gets picked out: Although we identify objects in other worlds by our own names, natives of some of these worlds use other names.

Second, as I have suggested in the text, 'Hesperus' is rigid because it picks out Hesperus in all worlds that *contain* Hesperus. In worlds not containing Hesperus, the designator fails to name anything other than Hesperus. Salmon (1980, pp. 32ff.) distinguishes between more than one understanding of a rigid designator that conforms to that requirement, but Kripke (1980, p. 21n.) deliberately ignores such complications.

Finally, even if (H) is not necessary, it will nevertheless be necessarily true of the object that is the brightest in the evening sky, Venus, that *it* is identical to Hesperus. Thus, Kripke's limited claim: "If '*a*' and '*b*' are rigid designators, it follows that '*a* = *b*', if true, is a necessary truth. If '*a*' and '*b*' are *not* rigid designators, no such conclusion follows about the *statement* '*a* = *b*'" (1980, p. 3, Kripke's emphasis).

2.  Recall that a clade is a species and all of its descendants. A clade stems from the closest common ancestor of all of its members apart from the stem itself. For more discussion about clades, including how ancestors *G* and *A* have been identified by systematists, refer to Chapter 1, §I.3. Note that the foregoing identities hold true *given* prominent systematists' use of 'Mammalia' and 'Aves'; again, see Chapter 1, §I.3.

3.  Putnam (1975e, p. 234) follows Kripke in affirming that kind terms are rigid. Putnam's explicit discussion of rigidity is brief, as he claims to express the same point at length in different terms (see also Putnam 1983a, pp. 57–8).

4.  Although there are no examples of scientifically discovered identities like the foregoing in the literature, Bolton does distinguish between theoretical identities and identities containing just names in an interesting and valuable discussion (1996, pp. 154–5).

    See Gould (1991, pp. 79ff.) for the story of '*Brontosaurus*'. Determining that the same species or genus has been named will occasionally be a matter of judgment, not discovery: The type specimens used to coin two species terms or genus terms may be members of the same species (or genus) by one plausible standard, but not by another plausible standard. But in some cases, it is clear that we have the same species (genus). The two terms may be grounded by type specimens that are conspecific by any reasonable measure: The terms may even be grounded by the same organism discovered twice, to take the most extreme case. The plurality of different standards for determining sameness of species, genus, and so on does not always ruin the claim that two organisms are discovered to belong to the same relevant taxon, then. It does, however, seriously undermine the claim that the standard that biologists happen to settle on reveals the discovered essence of the kind, as I argue in Chapter 3.

5.  There has been some controversy as to what makes kind terms rigid. For a quite different approach from the one I adopt here, see Cook (1980). It would be distracting for me to address Cook here, but Boër (1985, pp. 131ff.) provides illuminating criticisms. Readers interested in reviewing Cook's paper will find that my arguments about individual essences at the end of this chapter, if they are sound, also undermine his position. The position that kind designators should be taken to designate abstract entities rigidly has had its share of critics, whom I will address below, and its share of proponents: Boër (1985, pp. 129–35), Donnellan (1983, pp. 90f.), and

Mondadori (1978) have all taken kind terms to be "abstract nouns," as Donnellan puts it, following Mill (Donnellan 1983, p. 90).

6. See also Macbeth (1995, pp. 266, 268–71, 276–7). Both Schwartz and Macbeth discuss natural-kind terms as predicates here, rather than as singular terms. The intent of both is apparently to discuss natural-kind terms in general, however, so it appears that there is no intention to exclude from their observations kind terms used as singular terms. In any case, the issue *here* is natural-kind designators in general. Although I focus primarily on singular terms, similar words apply to predicates. Rather than discussing '*Brontosaurus*', or some other singular term for the same kind, like 'the brontosaur kind', I could discuss instead 'brontosaur', or '*member* of the brontosaur kind'. What follows could be adjusted accordingly.

7. Similar examples of nonrigid designators for natural kinds are offered by Sidelle (1989, pp. 62–3, esp. note 20) and also Nathan Salmon, who is acknowledged by Sidelle.

8. Putnam suggests this contrast (1975e, p. 265), as does, again, Schwartz (1980, p. 196). There may be some parallel between artificial-kind terms and descriptions of individuals, on the one hand, and natural-kind terms and proper names of individuals on the other, inasmuch as the *causal theory of reference* might apply to both members of the latter pair, and to neither member of the former pair (see §I.5 of this chapter). But these parallels do not reflect rigidity.

   For the sake of tidiness and readability, I have ignored certain complications here that I address elsewhere (LaPorte 2000): In particular, I have ignored the issue of whether the nonrigid designators I have discussed serve in some contexts to designate a second kind *rigidly* even though they serve in those contexts that I have been discussing to designate nonrigidly the kind that I take them to designate nonrigidly.

9. This has occasionally been disputed in the case of 'bachelor', but for the sake of sparing the popular foil, we can assume that tradition is right about this.

10. This view about rigidity's role is a common one. Gampel says that unlike 'water', 'can opener' is not rigid, because its reference is not tied "to actual samples of the kind" (Gampel 1997, p. 157). What makes 'water' rigid is that it refers to an underlying essence, not to whatever performs some function knowable a priori. Levine writes that rigid designators are "terms that refer independently of the truth of any descriptions associated with them" (2001, p. 328). De Sousa writes, "Rigid designators allow us to believe in essences without having to know what they are" (de Sousa 1984, p. 566). Many other writers express similar views about the job rigidity performs.

11. Perhaps it should be noted, to quell any doubts about the matter, that Putnam intends to apply 'rigid', following Kripke, to any term that "refers to the same individual in every possible world" (1975e, p. 231); he is not introducing a novel use for the term. Simple confusion about the relation between rigidity and the causal theory seems to explain why he believes that terms whose meaning is "in the head," like 'hunter', are not rigid, though causally grounded terms like 'water' are (see, besides the references immediately above, 1975e, p. 265). I am ignoring another possible confusion on Putnam's part, which is pointed out by Burge (1982, pp. 102–7): Causally grounded terms like 'water', says Burge, are not indexical at all.

12. An anonymous referee has done so.

13. This passage is from the *Essay*, III. ii. 2 (Locke 1975, p. 406). Locke explains his use of 'real essence' and 'nominal essence' in III. iii. 15 (p. 417). For an explicit passage about whether organisms' origins determine whether they belong in the extensions of biological kind terms, see III. vi. 23 (p. 452).

14. *P* must, of course, be independently specified, in order to save the definition from circularity. It would be specified by reference to a type specimen. A description that refers to *Panthera tigris*' type specimen would be used to fix the reference for '*P*', though it would not be synonymous with '*P*' (see the Introduction for a discussion of both type specimens and reference fixing). An example of such a description might be 'The earliest population whose descendants both (a) include such and such type specimen and (b) have not speciated before that type specimen lived'. As I have said, some independent account of speciation is needed, but this can be supplied by appeal to interbreeding or some other standard account.

15. It should be noted that not all of Hennig's followers endorse his claim that a species must become extinct when it splits; nevertheless, less strict cladistic concepts are subject to arguments very similar to the arguments given above (see Kitcher 1989, pp. 200–2, on Wiley 1981), as well as to the more general arguments below, which apply to both cladistic and noncladistic concepts.

## 3. BIOLOGICAL KIND TERM REFERENCE AND THE DISCOVERY OF ESSENCE

1. For the conclusion that guinea pigs are not rodents, see D'Erchia et al. (1996), and Graur, Hide, and Li (1991). The conclusion is highly controversial (see, e.g., Philippe 1997), but this can be ignored.

2. The idea that essentialism about biological kinds need only be reconceived along historical lines, and not abandoned in light of post–Darwinian systematics, has occurred to many writers; see also Matthen (1998, pp. 120–1, esp. note 24); de Queiroz (1992, p. 307); Ruse (1987, p. 236); Reimer (1997, pp. 38–9); Okasha (2002). Despite these authors, most of whose treatments are very brief, a great many philosophers of biology reject essentialism in all of its forms: This seems to me to be a mistake (see Chapter 1, §I.3, and Chapter 2).

   The recognition of historical essences is complicated by the fact that some maverick scientists do not grant that taxa are to be delimited historically: in particular, pheneticists and pattern cladists. In this chapter, I focus on the dominant position that taxa have historical essences, but the gist of what I have to say about the dominant position that taxa have historical essences could also be said about the minority position that taxa have nonhistorical essences. For simplicity, I will ignore minority voices.

3. I assume, for the sake of illustration, the validity of the conclusions drawn from the study mentioned above, ignoring controversy. I also ignore *non*cladistic standards for the present section. To broach noncladistic standards would reveal even more complications for the view that mistakes in the application of kind terms are discovered by scientists.

4. I use double quotation marks here because I am going back and forth in the use of key terms. Sometimes I am using terms like 'rodent' in the way scientists today use them. Sometimes I am using these terms as scientists *might* have used them. And

sometimes (in fact, most of the time I use quotation marks throughout the book) I am using terms as earlier, less scientifically sophisticated speakers used them.

5. On the cladistic view that birds are dinosaurs, see, for example, Dodson and Tatarinov (1990, p. 56). The claim now finds a place in many biology textbooks (see, e.g., Pough, Heiser, and McFarland 1996, pp. 405ff.).

6. See Humphries and Parenti (1986, pp. 22–4). Observe that even after scientists recognize a term to name no genuine taxon, they will often continue to use it, nevertheless, on account of its heritage. 'Algae' is a case in point (see, e.g., van den Hoek, Mann, and Jahns 1995, pp. 6–9).

7. Wiley (1981, p. 267) and Patterson (1993), for example, do not recognize the reptiles, because they gave rise to mammals and birds. This is the usual position, although some systematists continue to use 'reptile' for a monophyletic group by applying the word 'reptile' to animals that the traditional reptiles generated: As I have emphasized, these matters can go either way.

8. In fact, this *is* the interpretation of some biologists. Philippe (1997), for example, discusses the matter in terms of whether the rodents have been found to be non-monophyletic, in virtue of guinea pigs' ancestry. So if the supposed independent ancestry of guinea pigs is confirmed, then scientists may continue to call guinea pigs "rodents," taking 'rodent' to stand for a non-monophyletic group, or scientists may stop calling guinea pigs "rodents," taking 'rodent' to stand instead for the monophyletic group comprising of the other traditionally recognized rodents.

9. Occasionally people still do call whales "fish," as in telling the ancient story of the prophet Jonah, who was swallowed by a big fish (whale). But speakers will grant that strictly speaking we now understand that whales are not fish. Speakers are self-conscious about reserving 'fish' for relatively primitive, nonmammalian organisms. As it turns out, the fish do not constitute a natural group from a cladistic perspective, because advanced vertebrates evolved from them (see also the following note). This is not news, however, and cladists go on referring to whales as "nonfish" in full awareness of the artificial status of fish. Scientists have made a choice about how to use the term 'fish' in light of known information, and lay speakers have followed. I address lay use of vernacular terms more directly in section (IV) below.

10. 'Lizard' is discussed in Gauthier, Estes, and de Queiroz (1988, p. 16). Sometimes a term's use will be settled in the direction of scientific criteria in the face of one challenge, and nonscientific criteria in the face of another challenge. Thus, speakers decided not to call whales "fish" because they wanted to use 'fish' for a scientifically respectable group (gilled fish), but 'fish' has since been dropped from scientific use by cladists, who view it as an artificial-kind term. For cladists, the fish are an artificial group because not all descendants of fish are fish: Amphibians and then other more advanced vertebrates evolved from fish. In the face of this evolutionary challenge, cladists could still have opted to keep 'fish' tied to a scientifically respectable group by including more advanced vertebrates in the extension of 'fish', because the more advanced vertebrates evolved from the fish, but cladists did not choose to adjust the use of names in this way. Had they done so, whales would have turned out to belong to the extension of 'fish' after all, but not to the exclusion of the other mammals.

11. Of course, most scientists are not educated in philosophy, so scientists often do not use the terms 'essence' and 'accident', or have sophisticated views about essentialism. But essentialism follows from what scientists do say about what

makes members of the foregoing kinds members of those kinds. I discuss these matters at greater length in Chapter 1, §I.3, and in Chapter 2, §I.6 and §II.2.c.

12. This is a typical statement of the BSC. The concept has undergone changes over Mayr's many active decades. For a historical account of Mayr's work on the BSC, see Beurton (2002) and Mayr (2002).

13. Ridley, for example, repeatedly urges that the cladistic concept is "the correct definition" (1989, p. 1).

14. Martin reports on the meeting *Avian Taxonomy from Linnaeus to DNA*, London, March 23, 1996.

15. Again, see Martin (1996). As G. G. Simpson remarks, many taxonomists like to differentiate with a fine brush organisms in which they specialize, even though they hope that taxonomists working with organisms with which they are not so familiar will use a broad brush, which makes taxa handier for the nonspecialist (see Simpson 1961, p. 138).

16. However, some pluralistic accounts would not allow each strain of organisms to belong to two species. Mishler, Donoghue, and Brandon have all endorsed a pluralism according to which there is just one species to which any given organism belongs (Mishler and Donoghue 1982; Mishler and Brandon 1987). Their view is pluralist not because it allows organisms to belong to more than one species but rather because it allows that different species may be delimited variously: some by reproductive forces, others by ecological forces, still others by homeostatic inertia, and so on. Not all pluralists explicitly distinguish between these types of pluralism (see, e.g., Hey 2001, pp. 171–3).

17. A third major school, which I will ignore, is pheneticism (Sneath and Sokal 1973; Sokal and Sneath 1963). A nonhistorical school, pheneticism has become passé.

18. The foregoing examples and more are recorded in P. F. Stevens (1984; see esp. pp. 196ff.).

19. O'Brien (p.c.) is well aware of the optional nature of his decision. Notice that although the ranking of groups is largely arbitrary by all accounts (is this a subfamily? a family? a superfamily? a tribe?), this casts no doubt on the naturalness of ranked groups. The giant pandas do form a natural group, and so do the bears with or without the pandas.

20. For references, see Chapter 1, note 27.

21. The conversation was with James Walker, at the University of Massachusetts at Amherst, in the mid-1990s.

22. The theropods constitute a group of carnivorous dinosaurs that includes, for example, *Tyrannosaurus*. Birds are supposed to have evolved from the theropods. This quote is reported in Callahan (2001). For other media talk reflecting scientists' claim that birds are dinosaurs, see, for example, Gee (2000); Tangley (1997); and Spears (1993).

23. This is quoted in Spears (1993). My six-year-old son's book on dinosaurs (Branley 1989) teaches that some of the dinosaurs, rather than dying out, survived to become today's birds.

24. A minority of researchers still believes that the dinosaurs have not descended from the birds. If they are vindicated, reference to "avian dinosaurs" will of course dissipate.

25. Historical examples of this sort of response to surprising empirical findings are not easy to come by. One might argue that the discovery of marsupials resulted in a crisis for terms like 'wolf', which has been resolved by there being a scientific and a lay use for 'wolf': People call the marsupial wolf a "marsupial wolf," or "Tasmanian wolf." (Sometimes speakers also call it a "Tasmanian tiger," because of its stripes.) But there is not a lay and scientific use in this case. Even in scientific contexts, it is acceptable to use the term 'marsupial wolf', but both scientists and lay speakers grant that this is not a true wolf. A qualifier like 'marsupial' or 'Tasmanian' must not be ignored: Compare 'sea horse' or 'guinea pig', not to mention 'Tasmanian devil'. Strictly speaking, at least, a sea horse is not a true horse, and a Tasmanian wolf is not a true wolf (or tiger).

26. Lay speakers do not often seem to ignore scientists' use. Sometimes terms may have more than one sense, though. Dennett (1994, p. 535) argues that because tomatoes are "fruits" in the language of botany, there must be two definitions of 'fruit'. There is a strong case to be made for the claim. According to the botanical definition of 'fruit', cucumbers, green beans, peas, peppers, and many other such things count as "fruits." They are not "vegetables," which come from vegetative, or nonreproductive, organs. So perhaps 'fruit' has more than one sense. That is not to concede that there is any neat divide between vernacular and scientific use here. The U.S. Department of Agriculture, for example, which publishes the famous food pyramid found on food packaging, explicitly counts peas, beans, tomatoes, and the like to be vegetables when it recommends 3–5 vegetables a day. Scientists write the USDA's guidelines. These are experts (presumably not botanists but dieticians), not ordinary lay speakers.

27. Lay speakers' use of 'dinosaur' and other terms, and scientists' use as well, *is* context sensitive, though this does not necessarily indicate that the relevant term is vague, or that it has multiple senses. If a scientist looks into a polluted river and says, "There are no more salmon," her statement may be true in the context, a context in which only salmon in the river are under consideration. Still, the English word 'salmon' does not have vague application to salmon outside that river, nor does 'salmon' have multiple senses. Something similar might hold for the claim 'There are no more dinosaurs', said in a context in which only non-avian dinosaurs are under consideration. Avian dinosaurs are still dinosaurs, but they are ignored in the context.

28. It should be clear, then, that the criticism I raise against Kripke and Putnam is not just that natural-kind terms are vague, but rather that scientists constantly redefine terms (perhaps defeasibly) in response to vagueness. Kripke (1980, pp. 115n., 136) acknowledges that there is vagueness, but he says nothing about whether speakers might refine the vagueness out of terms, thus undermining his claims about essence discovery and the absence of meaning change. He appears not to recognize the problem, as he says that ordinarily "vagueness doesn't matter in practice" (1980, p. 136). It is not surprising that Kripke should suggest that vagueness poses little threat to his positions on the absence of meaning change and on discovered necessity. After all, 'Hesperus' and 'Phosphorus' are vague, but that does not suggest that their meanings were changed when people discovered the truth of 'Hesperus = Phosphorus', or that this sentence was not discovered to be necessarily true. Similar words apply to '*Brontosaurus = Apatosaurus*' (see Chapter 2). But

theoretical identity statements, in which the '=' is typically flanked by a familiar kind term on one side and a scientific expression revealing the kind's essence on the other side, are different. Like Kripke, Putnam similarly seems not to appreciate the importance of vagueness for his claims.

## 4. CHEMICAL KIND TERM REFERENCE AND THE DISCOVERY OF ESSENCE

1. This is not to deny that the word 'yü' is sometimes used quite loosely, much as our words 'golden' and 'silver' are. Tableware is often called "silver" even when it is stainless steel, costume jewelry is called "silver," and so on. *True* silver is a particular element, and it is harder to come by than the items commonly called "silver." In the same way, although the Chinese sometimes use 'yü' loosely to designate any beautiful, jade-like stone, they distinguish between *true* yü and other materials. Thus, Hansford (1948) writes that from very early times 'yü' was often used for "jade and jade-like stones" (p. 14), but he also indicates that even in very early times the Chinese did not count just any jade-like stone as "true jade": Before the nineteenth century, nephrite "had been the 'true jade' of China from time immemorial" (p. 15). Gump (1962; see pp. 17–20) compares the looser uses of 'yü' to looser uses of our word 'golden'. Like Westerners, the Chinese also refer to gold, of course, and again, authenticity is obviously just as important for the Chinese as it is for Westerners: You could not make a fortune by hauling fool's gold to China and passing it off as gold.

2. For the story of the introduction of jadeite into China, see especially Ward and Ward (1996, pp. 9–25, esp. 23–4); Gump (1962, pp. 180–91); and Hansford (1948, pp. 15–17).

   Note that, despite some critics' charge that the implausibility of the physical details of Putnam's XYZ example undermines his whole point (e.g., Brown 2002), the case of jade shows that at worst, *this* alleged problem could be circumvented by a change of examples: It is physically possible for substances to differ substantially in microstructure despite having similar observable properties, as even Putnam's remarks about jade indicate.

3. Ward and Ward (1996, p. 24) discuss 'new jade' and 'old jade'. Whitlock and Ehrmann (1949, p. 26) and Desautels (1986, p. 7) discuss compound names for colors of nephrite jade. Hansford (1948) discusses the history of the name 'kingfisher jade'.

4. Open texture is often said to be "the possibility of vagueness." I prefer to say that open texture is *actual* vagueness, not just possible vagueness, but that it is vagueness that is not known to characterize the application of a term to any actual candidates for reference. Both of the foregoing analyses are suggested by Waismann (1945, p. 123), who does not distinguish between them.

5. The disposition was lasting. Apparently the Chinese today have no regrets about having accorded jadeite the honor formerly reserved only for nephrite. One authority writes: "In hundreds of interviews inside and outside China and dozens of books, I have never encountered any reluctance at any level of society to converting from nephrite to jadeite" (Ward and Ward 1996, p. 24). Nevertheless, as I have said, the new stone does not seem to have been highly favored at first, at least if one writer present at the time of its introduction speaks for many. "I recollect

that, when I was young [kingfisher] jade of Yün-nan was not regarded as jade, but as merely usurping the name of jade," reminisces the statesman and scholar Chi Yün in a publication of 1800 (quoted in Hansford 1948, p. 17). Chi goes on to marvel at jadeite's turnaround: By the time of Chi's writing, jadeite carried a price tag exceeding that of the older nephrite. Chi's remark is curious because of its date: It was written no later than 1793 (according to the translator David Keenan, p.c.), when Chi was in his late sixties. Chi was writing about a period many years earlier ("when I was young"), which would have been long before the first large shipment of jadeite from Burma came to China, in 1784 (see esp. Gump 1962, pp. 181–2). When Chi was young, jadeite would have been obscure indeed, though it may have been present in very small quantities. One nineteenth-century British official relates a bit of folklore according to which a small Yünnanese trader introduced jadeite to China in the thirteenth century, but there is little evidence that the Chinese knew about jadeite at all before the eighteenth century: There are no extant jadeite carvings from earlier centuries, and there is no reliable testimony of there having been any of the material (Hansford 1948, p. 16; Gump 1962, pp. 182–4). Chi may, then, have misidentified the material from his youth. In any case, Chi had to have been talking about a period when jadeite was still little known and little understood, if it was around at all. Jadeite would rise from its obscurity quickly when the shipment of boulders to China in 1784 made its way to the emperor's studios. Because of its favor with the emperor, jadeite would come to be called "emperor's jade," as well as "kingfisher jade."

6.   Like carbon, $H_2O$ takes different forms, depending on temperature and pressure. $H_2O$ can take the solid form of ice, for instance. But this has long been a familiar manifestation of the stuff: Everyone has seen liquid freeze and then melt again. That is probably why people have little trouble saying that ice is frozen water. It is a less familiar fact that one can turn carbon in another state into diamond and back again by adjusting temperature and pressure. Whether people would call unusual $H_2O$ "water" may have to do with how easily convertible it is into one of the familiar states.

7.   One might complain that even though 'topaz' and 'water' have fairly natural *extensions*, they are not natural-kind terms, because it is unclear whether they refer to different materials with the same superficial properties in counterfactual worlds. But these terms can be called "natural-kind terms" in contexts that are rather relaxed (see Chapter 1, §II.3). Not only is there no XYZ in the actual world, so that 'water' has at minimum a relatively natural *extension*, but there is *some* disposition to say that XYZ in *other* possible worlds is "non-water." So a vernacular term like 'water' approximates to some degree the use of '$H_2O$', with respect to referring to the relevant natural kind.

8.   A standard encyclopedia's entry for "water" will confirm that $D_2O$ is considered water. Deuterium is a form of hydrogen and therefore one variety of $H_2O$. Therefore, if $D_2O$ were *not* water there would be trouble for the view that water = $H_2O$. Some philosophers seem to think, mistakenly, that deuterium is something other than hydrogen – for example, Churchland (1985, p. 4). Perhaps that is because deuterium is signified by the letter 'D' in '$D_2O$'. But deuterium is a form of hydrogen, not another element.

9. Donnellan (1983, pp. 97–104) seems to suppose otherwise, and this is a weakness in his case, which finds general agreement with the one I present in this section, despite some important differences. Donnellan bases his case on the dubious assertion that isotope number and atomic number vie for greater importance. Besides this mistake, Donnellan's case suffers from a paucity of empirical facts. The sketchiness of his empirical information leaves the reader with doubts about the plausibility of his claim that the physical world allows room for "wobbles of extension." Isotope variations had been declared inconsequential to reference before Donnellan (1983) (cf., e.g., Mackie 1976, p. 93), and a more-filled-in account than he offers is required to refute that claim. Despite these drawbacks, Donnellan's case shows delightful perception and ingenuity. Li (1993, pp. 272ff.) also argues that the extension of 'water' is indeterminate: I am not convinced by his particular argument, in part because it depends on XYZ and water's being microstructurally related, while Putnam clearly intends the contrary.

10. For example, with jade (more precisely, nephrite jade): This has exactly the same chemical and mineralogical composition as the minerals *actinolite* and *tremolite*. 'Jade' refers not to all matter with the relevant chemical/mineralogical composition but only to that with a distinct *fibrous structure*, which results in increased hardness.

11. Although in his classic writings (esp. 1975e) Putnam claims that it is metaphysically necessary that all and only $H_2O$ is water, the later Putnam rejects the whole notion of metaphysical possibility and necessity (1992, p. 443). Here and throughout I am obviously responding to the classic picture, with which Putnam is usually associated. I have little to say about Putnam's later turns of mood, as this is not an investigation of the excesses of Putnam's career.

## 5. LINGUISTIC CHANGE AND INCOMMENSURABILITY

1. The previous chapter might be thought to dilute the causal theory of reference, because I argue that although scientifically significant relations to samples help to determine reference, they do not trump superficial properties in determining reference; superficial properties play a major role, too. Still, the causal theory fails to secure stability even without the alterations to the theory that I recommend, as section (I) below indicates. In the same way, although I argue in the Introduction that some descriptive information is used in grounding terms even according to the causal theory, this role for descriptions is not the source of the instability that I describe in section (I). Because changes in the descriptive information associated with causally grounded words do not cause the referential instability that I discuss in (I), a purer version of the causal theory (see, e.g., Miller 1992) that dispenses with the role I have given to descriptive information does not evade the instability. The instability arises on account of features that are fundamental to the causal theory, rather than features that are limited to my preferred version of the causal theory.

2. For another explicit statement, see Kuhn (1983a, p. 684n.). For the later Kuhn, "the central characteristic of scientific revolutions" is the linguistic change they effect (1981, p. 28; see also p. 25, and 1979, p. 417). See section (IV) this chapter for more details on Kuhn's considered position.

3. For more discussion of these theories of reference, see the Introduction.

4. Both Kuhn and Feyerabend convey the foregoing message repeatedly and with many examples, just a few of which involve biology, the primary area of interest here. See, for example, Kuhn (1970a, pp. 128–9, 149, and the often-cited 102), and Feyerabend (1981a, pp. 44–5, 82–6, and throughout; 1981b; 1988).
5. Kuhn and Feyerabend have often been criticized for espousing relativism (see, e.g., Shapere 1981; McGrew 1994), though Kuhn denies the charge (Kuhn 1970b).
6. For references, see for example, Sankey (1994, p. 70, note 18). In this characterization of the problem, I have not located the difficulty with either meaning or reference in particular. Some have urged that the real issue is reference, not meaning; for example, Scheffler (1967, pp. 58–61) argues that meaning change in the absence of reference change does not give rise to incommensurability. I will set this issue aside by concentrating on language change affecting both meaning and reference. Kuhn and Feyerabend both recognize reference change as well as meaning change, as Kuhn points out in response to Scheffler (Kuhn 1970b, p. 269n.; see also 1981, p. 25). And the causal theory of reference is supposed to preserve sameness of reference as well as sameness of meaning (I am ignoring change in Putnam's stereotype, which could be counted as an irrelevant alteration of meaning).
7. Feyerabend is not explicit about his theory of reference, but according to a standard interpretation, the description theory of reference is the source of his extreme incommensurability (Sankey 1994, Chapter 5, e.g., pp. 144, 176 note 6; Devitt and Sterelny 1999, p. 228). This interpretation of Feyerabend is a natural one: The radical changes in reference that Feyerabend recognizes are apparently supposed to be triggered by changes in the descriptions associated with terms (see, e.g., 1981a, pp. 82–3). Kuhn sometimes suggests similarly radical reference change (see Fine 1975, p. 21; Sankey 1994, pp. 168, 172–3), but his general emphasis is more moderate.
8. Similar testimony is offered by Douven (2000, p. 136), Kornblith (1993, p. 6), Leplin (1988), and others. Putnam discusses incommensurability most explicitly in 1975d, for example, pp. 197–200, but he also discusses it in 1975e and 1975f, as well as Chapters 6 and 15 of 1975a.
9. For a discussion of how genetic information reveals ancestry, see for example O'Brien (1987), who explains how he has resolved the ancestry of a disputed group by using genetic clues.
10. This news may disappoint, but it is fortunate for the causal theory that it does *not* preclude meaning variance. Because language does change during scientific upheaval, any theory that cannot account for such change is in trouble (a point made by Nersessian 1991, p. 684).
11. As Kitcher (1983, p. 697) points out, linguistic change of the kind associated with incommensurability is actually not limited to revolutions but also attends relatively small advances. Think of Kuhn's own example of the platypus. Or think of examples from Chapter 3 involving the routine reconsideration of evolutionary relationships at various branches of the phylogenetic tree. Responding to this point, Kuhn (1983b, p. 714) concedes that he has become softer on the line between revolutions and "normal science" since writing *The Structure of Scientific Revolutions*. Even so, my focus here will be upon distinctively revolutionary change, because that is Kuhn's

focus and because such a focus highlights all that is at stake with respect to the issue of progress.

12. Letter to Ansted, 27 Oct. 1860, in Darwin (1903, Vol. 1, p. 175).

13. On Darwin's revision of the language see also Beatty (1985), whose excellent article has influenced the present section (II.1) of this chapter. Beatty traces Darwin's refinement to Watson.

14. The issue of how to use 'species' in conformity with pre-revolutionary use after the complete rejection of special creation was preceded by a similar issue that arose in the previous century after limited evolution from specially created forms was acknowledged. Linnaeus, for example, came to recognize that special creation could account for the birth of only a number of original stocks, but not for all of the distinct but interfertile groups into which each stock had apparently ramified. In the face of strong evidence that well-recognized "species" were somewhat interfertile, the question arose as to whether two or more interfertile groups formerly counted distinct "species" should be grouped together as a single "species." Linnaeus decided to continue to recognize the established "species." This forced him to forsake the sharp distinctions he had earlier recognized in his famous dictum "There are as many species as the Creator produced diverse forms in the beginning" (quoted in Glass 1968, p. 145). He retreated to a position that recognized the original creation of a limited number of forms that had later "speciated" (Glass 1968, pp. 145–61).

     Buffon, on the other hand, maintained that modification from original forms can result only in new "varieties," not in new "species." Still, the differences between Linnaeus and Buffon were, as Glass puts it, "purely semantic" (1968, p. 161). For Buffon, sterility is the test by which common ancestry from a given form was to be determined. Thus, Buffon allowed great modifications to occur within a "species": The fox and the dog could, for instance, be shown to be "conspecific" if tests of interfertility ever proved positive. Similarly, the zebra and ass or the zebra and horse might constitute a "species," as might the various apes (Lovejoy 1968, pp. 93, 95). Because little experimentation on sterility had been attempted, many questions about "conspecificity" could not be answered definitively, but Buffon did reduce substantially the number of "species" from the number Linnaeus had recognized.

15. In 1970a, p. 94, for example, Kuhn says that the arguments made on behalf of opposing theories structured by competing paradigms are circular. A Kuhnian paradigm is, roughly, a research tradition that presupposes certain beliefs, model problems, and model solutions, all of which provide a framework for further inquiry. After criticism, Kuhn has come to concede that he has used the term 'paradigm' in more than one distinct way.

16. Post-revolutionary statements may, of course, turn out to be vague, too. Just as earlier speakers' use of 'species' was rendered vague by evolution, the current use of 'species' may, for all we can be sure, be rendered vague by some presently unknown fact of revolutionary significance, so that 'New species arise by evolution' *still* fails to be straightforwardly true. For simplicity I will assume there is no such unknown trouble for the use of 'species'.

17. The possibility of retreating to a neutral language has been discussed by many philosophers, including Kuhn himself, who has reservations (Kuhn 1983a, pp. 671, 681–3; cf., e.g., Kitcher 1993, p. 275; Bird 1998, pp. 279–80). Kuhn's main concern seems to be, roughly, that because different speakers of a language identify a

term's referents by different means (e.g., children may distinguish the referents of 'man' and 'woman' by clothing rather than sex characteristics), the change in a few core theoretical terms may alter the way some speakers but not others pick out less-central terms' referents: Presumably some speakers but not others are supposed to depend upon the theoretical terms to pick out the referents of the less-central terms. But this seems to exaggerate linguistic change. Consider the sentence 'Darwin shares an ancestor with Rover'. It is very hard to believe that the altered use of theoretical terms like 'species' caused by evolutionism affects in the least how *anyone* identifies the referent of 'Darwin' or 'Rover'. Examples in the text seem similar, showing Kuhn's resistance to be wrongheaded (see also Kuhn 1970a, pp. 201–3).

18. Here, as throughout, I am assuming Kripke's familiar qualification: "*given* that scientists are right about how things are." I do not thereby assume what I set out to show, because the aim is not to show that scientists are right about how things are (this much I have assumed) but rather to show how scientists can have progressed in understanding the way things are in spite of the linguistic instability attending terms like 'species'. In the unlikely event that special creationists are right, of course, then the progress I discuss is chimerical, but even in such an event the example still illustrates how progress *could* occur despite linguistic instability.

    Because evolutionism can be expressed in pre-revolutionary English, a shift in terminological use is not essential to revolutionary change. Kuhn sometimes suggests the contrary, writing in one place about language change that "Until those changes had occurred, language itself resisted the invention and introduction of the sought after new theories" (1981, p. 28). Evolutionists could have abandoned affected terms and expressed their position that species evolve in other terms available in the language. Chambers dispenses with the term 'species' here (but compare the first edition of the *Vestiges*) rather than change its use. See also Hopkins, who shows that 'species' is dispensable, despite his continued use of the term (Hopkins 1973, p. 242).

19. See the range of views on vagueness in Keefe and Smith (1997a). The exceptional account would be the epistemic account of vagueness, on which vague sentences are just true or just false, period. But on such an account there is no telling which the sentence is, true or false, so again there is scientific progress in being able to replace a vague sentence with a nonvague one: If the replaced sentence is false, progress is made, and even if the replaced sentence is true, its successor is, unlike the sentence it supplants, *known* to be true.

20. Alternatively, the position that the relevant statement was meaningless might be combined with the view that the sentence was vague, rather than set in opposition to it. I do not commit to any particular account of the truth value of vague statements: whether they are true to a degree, or indeterminate, and so on. Hence, I do not rule out the possibility that vague statements have no truth value, either because they are meaningless or for some other reason. I thank Jim Allis and Andrew Dell'Olio in particular for their discussion of these issues.

21. Respondents who have offered feedback along these lines include an anonymous referee, whose worry I closely paraphrase in the text below.

22. At least they have failed to make progress of the sort that interests me here. Still, as Francis Bacon says, truth emerges more readily from error than from confusion.

This calls to mind Darwin's explicit frustration with miscommunication, which suggests that he would have been better off discussing his position with an audience that was just wrong about evolution but that spoke his refined language than with an audience whose unrefined language prevented its members even from communicating positions that were straightforwardly false.

23. Kuhn's cluster-of-descriptions account is no barrier to his position that there is common ground between succeeding theories, I have argued; on the contrary, it allows for common ground and is thus superior to Feyerabend's more radical alternative. Nevertheless, Kuhn frequently suggests that there is little common ground, as when he says that there are only circular arguments in favor of theories structured by competing paradigms, or when he says that scientists do not "zero in" over time on the truth about the world, or when he says that speakers at different times work in different "worlds." Certainly Kuhn does little to help the reader see where any common ground might lie.

24. Again, Kuhn seems to want to recognize disagreement, and he seems committed to it when he claims that successive theories are "incompatible" with one another (1970a, pp. 6, 97ff.). But again, some of his claims raise questions about whether he is entitled to these claims about incompatibility or to recognize disagreement (see, e.g., Logue 1995, p. 397).

25. Kuhn devotes Chapter 11 of 1970a to the topic of the rewriting of history; see also pp. 166–7.

26. This is clumsily stated, because the intent is clearly to count as vitalists those who say that any organism's vivacity *can* be explained "in terms of its form and composition" but that the composition of any organism is a mixture of an immaterial substance and chemical elements. The general characterization of vitalism given by Medawar and Medawar is ubiquitous. It is found also in, for example, Mayr (1982, p. 97) and Beckner (1967).

27. This is overstated even if vitalism is well summarized by the foregoing doctrines. Scientists have not *refuted* the claim that organisms have nonmaterial substances as parts; scientists rather ignore the idea as unfruitful or occult or unscientific. Nevertheless, Mayr is right that the foregoing vitalist theses are no longer taken seriously. And *chemical* vitalism *has* been refuted to every scientist's satisfaction, though textbooks commonly oversimplify the history, stating that Friedrich Wöhler laid the doctrine to rest in 1828 when he produced urea in the lab (see Cohen and Cohen 1996, p. 883). This is not true. For starters, many chemical vitalists denied that Wöhler had produced a genuinely organic chemical. The great Berzelius, for example, concluded from Wöhler's results that urea must not be a full-blooded organic compound (Teich and Needham 1992, p. 452). Further, the artificial synthesis of one organic chemical from inorganic ingredients does not go far in refuting a weaker version of chemical vitalism according to which only *some* organic chemicals could be so produced: Thus, decades after Wöhler's experiment, Pasteur defended the position that living organisms are required in fermentation.

28. These quotes are from Bernard (1957, p. 95). Bernard's claim to be a "physical vitalist" is discussed in, for example, Olmsted and Olmsted (1952, p. 214), and Goodfield (1960, p. 162): Goodfield's Chapters 7 and 8 helpfully discuss both Bernard and Liebig in connection with vitalism.

29. Schultz also discusses the political effects of vitalism in the dim reception of Galvani's work on muscle contraction. Goodfield (1960, pp. 146–7) suggests that Liebig's preoccupation with vital force may have deprived him of kudos he deserved for work on the conservation of energy.

30. See also Khalidi (1999) for a very good discussion of various interpretations.

31. Shapere, for example, adopts this interpretation (1981, p. 55; 1998, p. 733), as does Bird (1998, p. 277); cf. Lipton (1999, p. 164). Kuhn, at least in later writings, explicitly rejects this characterization. He makes many comparisons between theories, he points out: "[T]he book on which this interpretation is imposed includes many explicit examples of comparisons between successive theories" (1979, p. 416; see also 1983a, p. 670). This is true, but arguably not germane. Although Kuhn is surely right that he compares theories, he also suggests in many places that scientists wedded to one theory fail to take issue fully with competing theories. This suggests that comparison of the sort needed for rational discrimination between competing theories may be absent. And other sorts of comparison are irrelevant, as Feyerabend indicates. "Of course, *some* kind of comparison is *always* possible (for example, one physical theory may sound more melodious when read aloud to the accompaniment of a guitar than another physical theory)" (1988, p. 179). As one might expect, Kuhn is characteristically ambivalent on the issue, elsewhere conceding only "significant (though not complete) comparison" of rival theories (1983b, p. 713).

32. There are ancient and well-known difficulties with negative existential statements, to be sure, but these have little directly to do with problems of translation raised by incommensurability theorists. For a good review of the difficulties, see for example, Sainsbury (1991, Chapter 4).

One possible objection to my affirmation of the possibility of coining a term for a theoretical entity not recognized is this: New terms signifying entities that the dubber's theory rejects must thereby be terms of a *new* language, not the same theoretical language. This seems to multiply languages needlessly, but if you insist, I could adopt this way of delineating languages and urge that in that case, no scientist speaks just the language of her theory. A scientist speaks as many languages as there are theories she considers. Indeed, because we all consider various conflicting views, we all speak myriad languages.

33. Fine (1975, p. 18) attributes the use of this apt metaphor to N. R. Hanson.

34. Kuhn himself endorses enrichment, provided the terms from abandoned theories are "reserved for the special purposes of the philosopher, the historian, the writer of certain sorts of fiction" (1990, p. 308) and provided participants in the discourse keep tabs on whether current or obsolete uses are in effect. In this case, reasoning can proceed using older concepts but, as Kuhn indicates, "The question of translation simply does not arise" (1983a, p. 677). As usual, Kuhn is ambivalent on the subject, writing elsewhere that an attempt to enrich French by adding 'sweet' to it "would change preexisting distance relations and thus alter, rather than simply extend" the rest of the language (1983b, p. 714). This is hard to believe. It is common for bilingual speakers to complain about a lack of resources in one language and to remedy the situation by saying something like "This can only be said to be *doux*." This does not mangle English, so why would enrichment, which brings only more such occasions? In any case, cross-linguistic

communication is always a possibility by virtue of periodic borrowing, if not enrichment.

### 6. MEANING CHANGE, THEORY CHANGE, AND ANALYTICITY

1. That Quine's position on translation is widely rejected has been noted by many (see Gibson 1998, p. 28; Boghossian 1999, pp. 331–2; 1996, pp. 360–1; Lycan 1994, p. 249). Miller (1998, pp. 128–51) gives an overview of prominent animadversions in the literature. A great many philosophers deny the distinction between meaning change and theory change despite objections to Quine's views on translation (see the papers by Boghossian and Lycan).

2. Many earlier speakers did not measure angles by degrees, though. Euclid expresses the idea (*Elements* I.32) by saying that the sum of angles is equal to the sum of two right angles.

3. The great popularity of the thesis owes much not only to Quine but also to people like Harman (see also Harman 1967; 1996, pp. 397f.) and Putnam (1975e, pp. 254–6; 1975b), who have articulated the position with great power, using many plausible examples. Many others have made valuable contributions to the Quinean case; see, for example, Giaquinto (1996).

4. See, for example, Maor (1991, Chapter 16). The revolutionary advent of non-Euclidean geometry is discussed in many texts – for example, Dunham (1990, Chapter 2); Davis and Hersh (1980); Joyce (1998: see especially comments on Euclid's I.16).

5. In places, Putnam seems to deny that there has been meaning change: He says that "we should not say that 'straight line' has changed its meaning" (1975b, p. 50). Besides being wrong, this denial is problematic because it stands in obvious tension with Putnam's claim that meaning change and theory change are inseparable in cases like this (1975b, pp. 40–1; 1975e, p. 255).

6. Putnam (1975b) and Kitcher (1993, pp. 95–105) similarly recognize the falling apart of different conditions for reference in their admirable analyses of conceptual change. A related account is Field's classic (1973), which recognizes vagueness, or "partial reference," as I do. Others who recognize vagueness include Fine (1975, pp. 26–30), Boyd (1979, pp. 393ff.), and Papineau (1996). See also Shapere (1981, pp. 55ff.). Of course, these accounts are different from one another (and from my account) in various respects.

7. 'Big' and 'nice' are discussed in Keefe and Smith (1997b, p. 5). I borrow the discussion of 'religion' from Alston (1967). The terminology used to label the vagueness at issue varies. It is sometimes called "family resemblance vagueness," "cluster vagueness," "combinatory vagueness," and so on.

8. See the variety of views on vagueness in Keefe and Smith (1997a).

9. Darwin's spare definition of 'species' has since been explicitly augmented with more theory, of course, as Chapter 3 indicates, but now the theory reflects evolution. Beatty (1985) observes that like Darwin, Watson also refined the use of 'species' by prizing apart different conditions on reference generally attached to 'species'.

10. At least I would take that to be the case for the use of the term by the community as a whole. The community's use of 'species' depends on the conclusions of many

individuals and on a division of labor resulting in speakers' deference to others in the community. Though each individual is free to specify her own unique use for any term (Reid 2002), individuals ordinarily defer to the community's use of a term. Writers who stipulate idiosyncratic uses for words vary, of course, in how much their use is affected by others' use and in how much it would be affected by unexpected theoretical findings like evolution.

I have focused on biology. For a particularly explicit call from practicing scientists for the rejection of an orthodox "definition" in mineralogy after conditions have fallen apart, see Dietrich and Chamberlain (1989).

11. To those philosophers who take the minority position on this issue by resisting the case for necessity, I would submit the following argument as an argument to the effect that anyone who *does* accept the dominant position is committed to analyticity.

12. It is clear from passages like this that Quine is not arguing merely that there are borderline cases of analyticity. Nor is Quine merely saying that although there are perhaps a few analytic sentences, no doubt artificial ones resulting from stipulation, there are not many such sentences: He is arguing something much stronger. Indeed, for Quine analyticity seems to be incoherent or, as Quine could not say, coherent but *necessarily* uninstantiated (Boghossian 1999, pp. 340ff.). A more Quinean construal of the point is provided by Harman, in a justly celebrated discussion of Quine on analyticity: "[T]he analytic–synthetic distinction does not resemble the red–orange distinction, which is a distinction although a vague one. It resembles rather the witch–nonwitch distinction, which fails to distinguish anything since there are no witches" (1967, p. 125).

13. Here I am talking about a statement that has the variety of analyticity in question: a statement that is necessarily true in sole virtue of synonymy and logical truth. From now on, I will assume the qualification.

14. Similar words apply to apriority. There is no reason an apriorist should be embarrassed by fallibilism: Mathematical beliefs are sometimes false, but that is not much of an argument against their apriority. Not even the epistemic possibility of specifically *empirical* refutation should trouble the apriorist. Even if any truth might, for all we are guaranteed against the skeptic, be rationally *relinquished* on the basis of empirically acquired information – for example, on the basis of testimony – some knowledge may still be rationally *acquired* without the need for empirical investigation. See, in addition to Sober (pp. 64–7), Plantinga's helpful discussion of the matter (1993, pp. 110–13; also pp. 18–19).

Georges Rey (1993, pp. 72, 81, and elsewhere) and Putnam (1983b, pp. 96–7) emphasize that analyticity does not require infallibility, but many students of analyticity say that analytic statements are infallibly held or that they could not be rationally surrendered (Putnam 1975b, p. 54; Ayer 1952, pp. 72–7). Critics have rightly objected that no allegedly analytic statement seems to meet this standard (see, e.g., Lycan 1994, pp. 254–5).

15. Proponents of the new, causal theory of reference *can*, then, join Quine in rejecting the position that logic is analytic in any nontrivial respect (Quine 1976a; 1976b). In "Two Dogmas" Quine concedes, of course, the *trivial* analyticity of logical truth. There he takes aim at any further analyticity to be generated by synonymy. *Here* the proponent of the new theory of reference must resist.

16. Here I assume that a necessarily true sentence is straightforwardly true in all possible worlds. See note 25.

17. If the analyticity in question is a priori, as I suppose, then this analyticity will do *further* epistemological work: It will help to explain some a priori knowledge. As Boghossian says, if "a significant range of a priori statements are Frege-analytic, then the problem of their apriority is *reduced* to that of the apriority of logic and synonymy and, in this way, a significant economy in explanatory burden is achieved" (1999, p. 339. Boghossian presents a generally insightful discussion of nontrivial Frege-analyticity, or analyticity of Quine's second type, which is roughly the type at issue here; see Frege 1953, §3). I concur with the usual view that knowledge of analyticity would be a priori in part because I take facts about synonymy (at least the synonymy at issue) to be knowable a priori, but I will not defend this position here; cf. Devitt and Sterelny (1999, pp. 103, 179–84) and Margolis and Laurence (2001, p. 300).

18. Thus, Kripke found himself, in penning "Identity and Necessity," addressing a philosophical community that used 'analytic', 'necessary', and other purportedly equivalent terms interchangeably. Not everyone had use for such terms. Some, Kripke reports, "are vociferous defenders of them, and others, such as Quine, say they are all identically meaningless. But usually they're not distinguished" (1971, p. 149).

19. Many other philosophers testify to analyticity's poor reception. Lowe, for example, agrees that "Most contemporary philosophers are very wary of appearing to endorse the analytic–synthetic distinction following W. V. Quine's devastating onslaught upon it" (1995, p. 28). Similar testimony is offered by Ayer (1992, p. 479), Lycan (1994, p. 249), Bealer (1998, p. 234), Boghossian (1999), Margolis and Laurence (2001, p. 293), Friedman (2002, p. 177), and many others.

20. To be more precise, Kripke gives just this story when discussing 'Hesperus = Phosphorus' and 'Cicero = Tully'. He does not articulate in any detail how the account applies to statements about kinds, like '*Apatosaurus = Brontosaurus*', nor does he explicitly discuss examples like this. Kripke clearly thinks that statements about kinds are relevantly similar to statements about concrete objects, but I fill in gaps left by Kripke when I apply his account to statements like '*Apatosaurus = Brontosaurus*' (see Chapter 2 for details). If, for whatever reason, you have doubts about my application of Kripke's arguments for necessity to statements about kinds, this should not interfere with your reception of what follows. Simply substitute statements like 'Hesperus = Phosphorus' whenever I discuss statements like '*Apatosaurus = Brontosaurus*'. There is, of course, no need to discuss identity statements, either. Other statements would do as well, such as 'If Hesperus is bright, so is Phosphorus'.

21. Observe, too, that because '*Brontosaurus*' and '*Thunder-Lizardosaurus*' would be *rigid designators*, as well as synonyms, it follows that in addition to being analytic, '*Brontosaurus = Thunder-Lizardosaurus*' is, like '*Brontosaurus = Apatosaurus*' or 'Cicero = Tully', necessarily true in virtue of being an identity statement containing two rigid designators for the same thing.

22. It would be strange if it were not possible to generate synonymy by less formal means, too. Kripke emphasizes that although it is useful to imagine that terms are coined in artificial baptisms, causally grounded terms are commonly coined

by less formal means (Kripke 1980, p. 162). Yet if the coining of '*Brontosaurus*' and '*Apatosaurus*' as rigid designators for *Apatosaurus* (*Brontosaurus*) can be achieved informally rather than by a ceremonial baptism involving a pronouncement like "I coin '*Brontosaurus*' as a term for the genus instantiated in such and such sample," then one should expect, at least in the absence of cogent reasons for distinguishing the two cases – which hardly seem likely to be forthcoming – that the coining of '*Thunder-Lizardosaurus*' as a rigid designator for *Brontosaurus* could also be achieved by a means less formal than the pronouncement "I coin '*Thunder-Lizardosaurus*' as a term for *Brontosaurus*."

23. See, for example, Margolis and Laurence (2001, p. 301) and Putnam (1975b, p. 45) on this frequently articulated point. (Putnam makes an exception for "one-criterion" words, as opposed to words from science; because scientific words are the ones at issue here, this exception may be ignored.)

24. See Putnam (1975e, pp. 255–6). Putnam is explicit that he is talking about metaphysical necessity, not epistemic possibility. A posteriori investigation is needed to reveal the necessary truth of 'Water is $H_2O$', so there is clearly an epistemic possibility that the statement is false and that empirical science has things wrong. I discuss the epistemic possibility of revision further below, when addressing the Quinean slogan 'There is no guarantee against revision for any statement'.

    In earlier work, Putnam qualifies his claim that the distinction between meaning change and theory change is spurious: He suggests that the distinction is spurious *unless* sentences contain "one-criterion" words (1975b). Because sentences like 'Water is $H_2O$' do not contain one-criterion words, the qualification is irrelevant here.

25. I assume here that a necessarily true statement is *straightforwardly* true in any possible world, so that partial falsity or falsity to a degree, and so on, owing to vagueness or confusion, is precluded. Otherwise, the metaphysical possibility of revision needed to iron out partial error would be compatible after all with necessity, including analytic necessity. Of course, even if a statement is straightforwardly true in any possible world and therefore metaphysically necessary, there is still an *epistemic* possibility that it or any other statement may turn out to be partially false or not straightforwardly true because of vagueness. The epistemic possibility of partial falsity is compatible with there being metaphysical necessity and analyticity for the same reason that the epistemic possibility of outright error is compatible with there being metaphysical necessity and analyticity.

# References

Abbott, Barbara. 1997. "A Note on the Nature of 'Water'." *Mind* 106: 311–19.

Alston, William. 1967. "Vagueness." In *The Encyclopedia of Philosophy*, vol. 8, ed. Paul Edwards, pp. 218–21. New York: Macmillan.

Andersson, L. 1990. "The Driving Force: Species Concepts and Ecology." *Taxon* 39: 375–82.

Angier, Natalie. 1996. "Guinea Pigs Not Rodents? DNA Weighing In." *New York Times*, 13 June.

Armstrong, D. M. 1983. *What Is a Law of Nature?* New York: Cambridge University Press.

Aune, Bruce. 1994. "Determinate Meaning and Analytic Truth." In *Living Doubt*, ed. G. Debrock and M. Hulswit, pp. 55–65. Netherlands: Kluwer Academic Publishers.

Ayer, A. J. 1952. *Language, Truth, and Logic*. Second Edition. New York: Dover.

1992. "Reply to F. Miró Quesada." In *The Philosophy of A. J. Ayer*, ed. L. E. Hahn, pp. 478–88. La Salle, Ill.: Open Court.

Barnett, David. 2000. "Is Water Necessarily Identical to H$_2$O?" *Philosophical Studies* 98: 99–112.

Bates, Henry Walter. 1910. *The Naturalist on the River Amazons*. New York: E. P. Dutton & Co.

Bealer, George. 1998. "Analyticity." In *Routledge Encyclopedia of Philosophy*, vol. 1, ed. Edward Craig, pp. 234–9. London: Routledge.

Beatty, John. 1985. "Speaking of Species: Darwin's Strategy." In *The Darwinian Heritage*, ed. David Kohn, pp. 265–81. Princeton, N.J.: Princeton University Press.

Beckner, Morton. 1967. "Vitalism." In *The Encyclopedia of Philosophy*, vol. 8, ed. Paul Edwards, pp. 253–6. New York: Macmillan.

Bernard, Claude. 1957. *An Introduction to the Study of Experimental Medicine*. Translated into English by Copley Green, with an Introduction by Lawrence J. Henderson and a Foreword by I. Bernard Cohen. New York: Dover. Originally published as *Introduction à l'étude de la Médecine expérimentale*. Paris: J. B. Baillière, 1865.

Beurton, Peter. 2002. "Ernst Mayr through Time on the Biological Species Concept – A Conceptual Analysis." *Theory in Biosciences* 121: 81–98.

Bicheno, J. E. 1826. "On Systems and Methods in Natural History." *Transactions of the Linnæan Society* 15: 479–96.

201

Bird, Alexander. 1998. *Philosophy of Science*. Montreal: McGill–Queen's University Press.

Boër, Steven. 1985. "Substance and Kind: Reflections on the New Theory of Reference." In *Analytical Philosophy in Comparative Perspective*, ed. B. K. Matilal and J. L. Shaw, pp. 103–50. Dordrecht: D. Reidel.

Boghossian, Paul A. 1996. "Analyticity Revisited." *Noûs* 30: 360–91.

1999. "Analyticity." In *A Companion to the Philosophy of Language*, ed. Bob Hale and Crispin Wright, pp. 331–68. Oxford: Blackwell.

Bolton, Cynthia J. 1996. "Proper Names, Taxonomic Names and Necessity." *The Philosophical Quarterly* 46: 145–57.

Bowler, Peter J. 1994. "Are the Arthropoda a Natural Group? An Episode in the History of Evolutionary Biology." *Journal of the History of Biology* 27: 177–213.

Borjesson, Gary. 1999. "Not for Their Own Sake: Species and the Riddle of Individuality." *Review of Metaphysics* 52: 867–96.

Boyd, Richard. 1979. "Metaphor and Theory Change: What Is 'Metaphor' a Metaphor For?" In *Metaphor and Thought*, ed. A. Ortony, pp. 356–408. Cambridge: Cambridge University Press.

1991. "Realism, Anti-Foundationalism, and the Enthusiasm for Natural Kinds." *Philosophical Studies* 61: 127–48.

Branley, Franklyn. 1989. *What Happened to the Dinosaurs?* New York: Crowell.

Broad, C. D. 1925. *The Mind and Its Place in Nature*. New York: Humanities Press.

Brower, Andrew. 2000. "Evolution Is Not a Necessary Assumption of Cladistics." *Cladistics* 16: 143–54.

Brown, James R. 2002. Review of *Thomas Kuhn*, by Alexander Bird. *British Journal for the Philosophy of Science* 53: 143–9.

Brummitt, Richard. 2002. "How to Chop Up a Tree." *Taxon* 51: 31–41.

Burge, Tyler. 1982. "Other Bodies." In *Thought and Object*, ed. A Woodfield, pp. 97–120. Oxford: Oxford University Press.

Bynum, W. F. 1981. "Vitalism." In *Dictionary of the History of Science*, ed. W. F. Bynum, E. Browne, and Roy Porter, pp. 439–40. Princeton, N.J.: Princeton University Press.

Callahan, Rick. 2001. "Scientists: Fossil Dinosaur Was Covered in Primitive Feathers." *Associated Press*, 25 April. LEXIS-NEXIS. 26 April 2001 <http://web.lexis-nexis.com/universe>.

Chambers, Robert. 1845. *Explanations: A Sequel to "Vestiges of the Natural History of Creation."* London: John Churchill. (Facsimile in *Vestiges of the Natural History of Creation and Other Evolutionary Writings*, ed. James A. Secord. Chicago: University of Chicago Press, 1994.)

Churchland, Paul. 1985. "Conceptual Progress and Word/World Relations: In Search of the Essence of Natural Kinds." *Canadian Journal of Philosophy* 15: 1–17.

Clark, Andrew M. 1993. *Hey's Mineral Index: Mineral Species, Varieties and Synonyms*. Third Edition. London: Chapman & Hall.

Cocchiarella, N. 1976. "On the Logic of Natural Kinds." *Philosophy of Science* 43: 202–22.

Cohen, Paul S., and Cohen, Stephen M. 1996. "Wöhler's Synthesis of Urea: How Do the Textbooks Report It?" *Journal of Chemical Education* 73: 883–6.

Coleman, William. 1977. *Biology in the Nineteenth Century: Problems of Form, Function, and Transformation*. Cambridge: Cambridge University Press.

Collier, John. 1996. "On the Necessity of Natural Kinds." In *Natural Kinds, Laws of Nature and Scientific Methodology*, ed. P. Riggs, pp. 1–10. Dordrecht: Kluwer Academic Publishers.

Collins, H. M. 1985. *Changing Order: Replication and Induction in Scientific Practice*. London: Sage.

Cook, Monte. 1980. "If 'Cat' Is a Rigid Designator, What Does It Designate?" *Philosophical Studies* 37: 61–4.

Cracraft, Joel. 1983. "Species Concepts and Speciation Analysis." In *Current Ornithology*, ed. R. F. Johnston, pp. 159–87. New York: Plenum Press.

Daly, Chris. 1998. "Natural Kinds." In *Routledge Encyclopedia of Philosophy*, vol. 6, ed. Edward Craig, pp. 682–5. London: Routledge.

Darwin, Charles. 1859. *On the Origin of Species*. London: John Murray. (Facsimile: Cambridge, Mass.: Harvard University Press, 1964.)

　　　1871. *The Descent of Man, and Selection in Relation to Sex*, 2 vols. London: John Murray. (Facsimile, 2 vols. in 1: Princeton, N.J.: Princeton University Press, 1981.)

　　　1903. *More Letters of Charles Darwin: A Record of His Work in a Series of Hitherto Unpublished Letters*, 2 vols., ed. Francis Darwin and A. C. Seward. New York: D. Appleton and Company.

　　　1975. *Charles Darwin's Natural Selection: Being the Second Part of His Big Species Book Written From 1856 to 1858*, ed. R. C. Stauffer. New York: Cambridge University Press.

Davis, Philip, and Hersh, Reuben. 1980. *The Mathematical Experience*. Boston: Birkhäuser.

Dennett, Daniel. 1994. "Get Real." *Philosophical Topics* 22: 505–68.

De Queiroz, Kevin. 1992. "Phylogenetic Definitions and Taxonomic Philosophy." *Biology and Philosophy* 7: 295–313.

　　　1994. "Replacement of an Essentialistic Perspective on Taxonomic Definitions as Exemplified by the Definition of 'Mammalia'." *Systematic Biology* 43: 497–510.

　　　1995. "The Definitions of Species and Clade Names: A Reply to Ghiselin." *Biology and Philosophy* 10: 223–8.

De Queiroz, Kevin, and Cantino, Philip. 2001. "Phylogenetic Nomenclature and the PhyloCode." *Bulletin of Zoological Nomenclature* 58: 254–71.

De Queiroz, Kevin, and Donoghue, Michael J. 1988. "Phylogenetic Systematics and the Species Problem." *Cladistics* 4: 317–38.

De Queiroz, Kevin, and Gauthier, Jacques. 1994. "Toward a Phylogenetic System of Biological Nomenclature." *Trends in Ecology and Evolution* 9: 27–31.

D'Erchia, Anna Maria, et al. 1996. "The Guinea-Pig Is Not a Rodent." *Nature* 381: 597–600.

De Sousa, Ronald. 1984. "The Natural Shiftiness of Natural Kinds." *Canadian Journal of Philosophy* 14: 561–80.

Desautels, Paul. 1986. *The Jade Kingdom*. New York: Van Nostrand Reinhold.

Devitt, Michael, and Sterelny, Kim. 1999. *Languages and Reality: An Introduction to the Philosophy of Language*. Second Edition. Cambridge, Mass.: MIT Press.

Dietrich, R. V., and Chamberlain, Steven. 1989. "Are Cultured Pearls Mineral? The Continuing Evolution of the Definition of *Mineral*." *Rocks & Minerals* 64: 386–92.

Dietrich, R. V., and Skinner, Brian J. 1990. *Gems, Granites, and Gravels*. Cambridge: Cambridge University Press.

Dobzhansky, Theodosius. 1970. *Genetics of the Evolutionary Process*. New York: Columbia University Press.

Dodson, Peter, and Tatarinov, Leonid P. 1990. "Dinosaur Extinction." In *The Dinosauria* ed. D. B. Weishampel, P. Dodson, and H. Osmólska, pp. 55–62. Berkeley and Los Angeles: University of California Press.

Doepke, F. 1992. "Identity and Natural Kinds." *The Philosophical Quarterly* 42: 89–94.

Donnellan, Keith. 1983. "Kripke and Putnam on Natural Kind Terms." In *Knowledge and Mind*, ed. C. Ginet and S. Shoemaker, pp. 84–104. New York: Oxford University Press.

Donoghue, Michael. 2001. "A Wish List for Systematic Biology." *Systematic Biology* 50: 755–7.

Douven, Igor. 2000. "Theoretical Terms and the Principle of the Benefit of Doubt." *International Studies in the Philosophy of Science* 14: 135–46.

Douven, Igor, and Van Brakel, Jaap. 1998. "Can the World Help Us in Fixing the Reference of Natural Kind Terms?" *Journal for General Philosophy of Science* 29: 59–70.

Dunham, William. 1990. *Journey Through Genius: The Great Theorems of Mathematics*. New York: Wiley.

Dupré, John. 1993. *The Disorder of Things: Metaphysical Foundations of the Disunity of Science*. Cambridge, Mass.: Harvard University Press.

1999. "Are Whales Fish?" In *Folkbiology*, ed. D. Medin and S. Atran, pp. 461–76. Cambridge, Mass.: MIT Press.

2001. "In Defence of Classification." *Studies in History and Philosophy of Science Part C: Studies in History and Philosophy of Biological and Biomedical Sciences* 32: 203–19.

Ellis, Brian. 2001. *Scientific Essentialism*. Cambridge: Cambridge University Press.

Emmeche, Claus, Køppe, Simo, and Stjernfelt, Frederik. 1997. "Explaining Emergence: Towards an Ontology of Levels." *Journal for General Philosophy of Science* 28: 83–119.

Ereshefsky, Marc. 1991. "Species, Higher Taxa, and the Units of Evolution." *Philosophy of Science* 58: 84–101.

1992. "Eliminative Pluralism." *Philosophy of Science* 59: 671–90.

2001. *The Poverty of the Linnaean Hierarchy: A Philosophical Study of Biological Taxonomy*. Cambridge: Cambridge University Press.

Fawcett, Henry. 1973. "A Popular Exposition of Mr. Darwin on the Origin of Species." Reprinted in *Darwin and His Critics: The Reception of Darwin's Theory of Evolution by the Scientific Community*, ed. David Hull, pp. 277–90. Cambridge, Mass.: Harvard University Press.

Feldmaier, Carl. 1970. *Lilies*. New York: Arco.

Feyerabend, Paul. 1978. *Science in a Free Society*. London: NLB.

1981a. "Explanation, Reduction, and Empiricism." In *Realism, Rationalism and Scientific Method: Philosophical Papers*, vol. 1, pp. 44–96. Cambridge: Cambridge University Press.

1981b. "On the 'Meaning' of Scientific Terms." In *Realism, Rationalism and Scientific Method: Philosophical Papers*, vol. 1, pp. 97–103. Cambridge: Cambridge University Press.

1988. *Against Method*. Revised Edition. London: Verso.

Field, Hartry. 1973. "Theory Change and the Indeterminacy of Reference." *Journal of Philosophy* 70: 462–81.

Fine, Arthur. 1975. "How to Compare Theories: Reference and Change." *Noûs* 9: 17–32.

Finn, Jane Adams. 1997. "Sweet on Onions: With Vidalias, Walla Wallas and Mauis, You'll Never Shed a Tear." *Washington Post*, 19 February. LEXIS-NEXIS. 15 June 2001 <http://web.lexis-nexis.com/universe>.

Forey, P. L. 2002. "*PhyloCode* – Pain, No Gain." *Taxon* 51: 43–54.

Frege, Gottlob. 1953. *The Foundations of Arithmetic*. Trans. J. L. Austin. Oxford: Basil Blackwell.

Friedman, Michael. 2002. "Kant, Kuhn, and the Rationality of Science." *Philosophy of Science* 69: 171–90.

Gampel, Eric H. 1997. "Ethics, Reference, and Natural Kinds." *Philosophical Papers* 26: 147–63.

Gams, W. 1995. "How Natural Should Anamorph Genera Be?" *Canadian Journal of Botany* 73: S747–S753.

"Garlic." *Encyclopædia Britannica Online*. 21 June 2001 <http://search.eb.com/eb/article?eu=36779>.

Gauthier, Jacques, Estes, Richard, and de Queiroz, Kevin. 1988. "A Phylogenetic Analysis of Lepidosauromorpha." In *Phylogenetic Relationships of the Lizard Families*, ed. R. Estes and G. Pregill, pp. 15–98. Stanford, Calif.: Stanford University Press.

Gee, Henry. 2000. "Science: Reptile Ruffles Feathers." *Guardian*, 7 December, Guardian Science Pages. LEXIS-NEXIS. April 26, 2001 <http://web.lexis-nexis.com/universe>.

Ghiselin, Michael. 1974. "A Radical Solution to the Species Problem." *Systematic Zoology* 23: 536–44.

1980. "Natural Kinds and Literary Accomplishments." *Michigan Quarterly Review* 19: 73–88.

1981. "Taxa, Life, and Thinking." *Behavioral and Brain Sciences* 4: 303–10.

1987. "Species Concepts, Individuality, and Objectivity." *Biology and Philosophy* 2: 127–43.

1989. "Individuality, History and Laws of Nature in Biology." In *What the Philosophy of Biology Is: Essays Dedicated to David Hull*, ed. M. Ruse, pp. 53–66. Dordrecht: Kluwer Academic Publishers.

1995. "Ostensive Definitions of the Names of Species and Clades." *Biology and Philosophy* 10: 219–22.

1997. *Metaphysics and the Origin of Species*. Albany, N.Y.: SUNY Press.

1999. "Natural Kinds and Supraorganismal Individuals." In *Folkbiology*, ed. D. Medin and S. Atran, pp. 447–60. Cambridge, Mass.: MIT Press.

Giaquinto, M. 1996. "Non-Analytic Conceptual Knowledge." *Mind* 105: 249–68.

Gibson, Roger. 1998. "Radical Translation and Radical Interpretation." In *Routledge Encyclopedia of Philosophy*, vol. 8, ed. Edward Craig, pp. 25–35. London: Routledge.

Glass, Bentley. 1968. "Heredity and Variation in the Eighteenth Century Concept of the Species." In *Forerunners of Darwin, 1745–1859*, ed. B. Glass, O. Temkin, and W. Straus, pp. 144–72. Baltimore: The Johns Hopkins Press.

Goldsmith, Oliver. 1791. *An History of the Earth and Animated Nature*, 8 vols. London: F. Wingrave.

Goodfield, G. J. 1960. *The Growth of Scientific Physiology*. London: Hutchinson.

Goodman, Nelson. 1955. *Fact, Fiction and Forecast*. Cambridge, Mass.: Harvard University Press.

Gould, Stephen J. 1983. "What, if Anything, Is a Zebra?" In *Hen's Teeth and Horse's Toes*, pp. 355–65. New York: Norton.

1991. *Bully for Brontosaurus: Reflections in Natural History*. New York: Norton.

Graur, Dan A., Hide, Winston, and Li, Wen-Hsiung. 1991. "Is the Guinea-Pig a Rodent?" *Nature* 351: 649–52.

Gregory, Frederick. 1977. *Scientific Materialism in Nineteenth Century Germany*. Dordrecht: D. Reidel.

Griffiths, P. E. 1994. "Cladistic Classification and Functional Explanation." *Philosophy of Science* 61: 206–27.

Gump, Richard. 1962. *Jade: Stone of Heaven*. Garden City, N.Y.: Doubleday.

Hacking, Ian. 1991. "A Tradition of Natural Kinds." *Philosophical Studies* 61: 109–26.

Hansford, S. Howard. 1948. "Jade and the Kingfisher." *Oriental Art* 1: 12–17.

Harman, Gilbert. 1967. "Quine on Meaning and Existence, I." *Review of Metaphysics* 21: 124–51.

1973. *Thought*. Princeton, N.J.: Princeton University Press.

1996. "Analyticity Regained?" *Noûs* 30: 392–400.

Hennig, Willi. 1965. "Phylogenetic Systematics." In *Annual Review of Entomology*, ed. R. Smith and T. Mittler, vol. 10, pp. 97–116. Palo Alto, Calif.: Annual Reviews, Inc.

1966. *Phylogenetic Systematics*. Urbana: University of Illinois Press.

Hey, Jody. 2001. *Genes, Categories, and Species: The Evolutionary and Cognitive Causes of the Species Problem*. New York: Oxford University Press.

Hopkins, William. 1973. "Physical Theories of the Phenomena of Life." Reprinted in *Darwin and His Critics: The Reception of Darwin's Theory of Evolution by the Scientific Community*, ed. David Hull, pp. 229–72. Cambridge, Mass.: Harvard University Press.

Hughes, Richard W. 1990. *Corundum*. London: Butterworth-Heinemann.

Hull, David. 1973. *Darwin and His Critics: The Reception of Darwin's Theory of Evolution by the Scientific Community*. Cambridge, Mass.: Harvard University Press.

1976. "Are Species Really Individuals?" *Systematic Zoology* 25: 174–91.

1978. "A Matter of Individuality." *Philosophy of Science* 45: 335–60.

1980. "Individuality and Selection." *Annual Review of Ecology and Systematics* 11: 311–32.

1981. "Metaphysics and Common Usage." *Behavioral and Brain Sciences* 4: 290–1.

1987. "Genealogical Actors in Ecological Roles." *Biology and Philosophy* 2: 168–84.

1988. *Science as a Process*. Chicago: University of Chicago Press.

1998. "Taxonomy." In *Routledge Encyclopedia of Philosophy*, vol. 9, ed. Edward Craig, pp. 272–6. London: Routledge.

2001. "The Role of Theories in Biological Systematics." *Studies in History and Philosophy of Science Part C: Studies in History and Philosophy of Biological and Biomedical Sciences* 32: 221–38.

Humphries, Christopher J., and Parenti, Lynne R. 1986. *Cladistic Biogeography*, pp. 22–4. Oxford: Clarendon Press.

Jackson, Peter. 2000. "Subspecies and Conservation." *Cat News* 32 (Spring). The Tiger Information Center. 1 August 2001 <http://www.5tigers.org/NewsLetters/CatNews/cn32/editorial.htm>.

Jacob, François. 1982. *The Logic of Life: A History of Heredity*. New York: Pantheon Books.

Johnson, L. A. S. 1970. "Rainbow's End: The Quest for an Optimal Taxonomy." *Systematic Zoology* 19: 203–39.

Joyce, David, ed. 1998. *Euclid's Elements*, by Euclid. 17 July 2001 <http://aleph0.clarku.edu/~djoyce/java/elements/elements.html>.

Keefe, Rosanna, and Smith, Peter. 1997a. *Vagueness: A Reader*. Cambridge, Mass.: MIT Press.

1997b. "Introduction: Theories of Vagueness." In *Vagueness: A Reader*, pp. 1–57. Cambridge, Mass.: MIT Press.

Keverne, Roger. 1991. *Jade*. London: Anness Publishing.

Khalidi, Muhammad Ali. 1999. "Incommensurability." In *A Companion to Philosophy of Science*, ed. W. H. Newton-Smith, pp. 172–80. Oxford: Blackwell.

Kitcher, Philip. 1978. "Theories, Theorists and Theoretical Change." *Philosophical Review* 87: 519–47.

1983. "Implications of Incommensurability." In *PSA 1982*, vol. 2, ed. P. D. Asquith and T. Nickles, pp. 689–703. East Lansing, Mich.: PSA.

1984. "Species." *Philosophy of Science* 51: 308–33.

1987. "Ghostly Whispers: Mayr, Ghiselin, and the 'Philosophers' on the Ontological Status of Species." *Biology and Philosophy* 2: 184–92.

1989. "Some Puzzles About Species." In *What the Philosophy of Biology Is: Essays Dedicated to David Hull*, ed. M. Ruse, pp. 183–208. Boston: Kluwer Academic Publishers.

1993. *The Advancement of Science: Science Without Legend, Objectivity Without Illusions*. Oxford: Oxford University Press.

Kornblith, Hilary. 1980. "Referring to Artifacts." *The Philosophical Review* 89: 109–14.

1993. *Inductive Inference and Its Natural Ground*. Cambridge, Mass.: MIT Press.

Kraft, James Lewis. 1947. *Adventure in Jade*. New York: Holt.

Kripke, Saul. 1971. "Identity and Necessity." In *Identity and Individuation*, ed. M. K. Munitz, pp. 135–64. New York: New York University Press.

1977. "Speaker's Reference and Semantic Reference." In *Midwest Studies in Philosophy, II, Studies in the Philosophy of Language*, ed. P. French, T. Uehling, and H. Wettstein, pp. 225–76. Morris: University of Minnesota Press.

1980. *Naming and Necessity*. Cambridge, Mass.: Harvard University Press.

Kuhn, Thomas S. 1970a. *The Structure of Scientific Revolutions*. Second Edition. Chicago: University of Chicago Press.

1970b. "Reflections on My Critics." In *Criticism and the Growth of Knowledge*, ed. Imre Lakatos and Alan Musgrave, pp. 231–78. Cambridge: Cambridge University Press.

1979. "Metaphor in Science." In *Metaphor and Thought*, ed. A. Ortony, pp. 409–19. Cambridge: Cambridge University Press.

1981. "What Are Scientific Revolutions?" Occasional Paper #18, Cambridge, Mass.: Center for Cognitive Science, MIT.

1983a. "Commensurability, Comparability, Communicability." In *PSA 1982*, vol. 2, ed. P. D. Asquith and T. Nickles, pp. 669–88. East Lansing, Mich.: PSA.

1983b. "Response to Commentaries." In *PSA 1982*, vol. 2, ed. P. D. Asquith and T. Nickles, pp. 712–16. East Lansing, Mich.: PSA.

1990. "Dubbing and Redubbing: The Vulnerability of Rigid Designation." In *Minnesota Studies in the Philosophy of Science*, vol. 14, ed. C. Wade Savage, pp. 298–318. Minneapolis: University of Minnesota Press.

2000a. "Possible Worlds in the History of Science." In *The Road Since Structure*, ed. James Conant and John Haugeland, pp. 58–89. Chicago: University of Chicago Press.

2000b. "Afterwords." In *The Road Since Structure*, ed. James Conant and John Haugeland, pp. 224–52. Chicago: University of Chicago Press.

Lange, Marc. 1995. "Are There Natural Laws Concerning Particular Biological Species?" *Journal of Philosophy* 92: 430–51.

LaPorte, J. 1996. "Chemical Kind Term Reference and the Discovery of Essence." *Noûs* 30: 112–32.

2000. "Rigidity and Kind." *Philosophical Studies* 97: 293–316.

forthcoming. Review of *Genes, Categories, and Species*, by Jody Hey. *British Journal for the Philosophy of Science*.

Leakey, Richard, ed. 1979. *The Illustrated Origin of Species*, by Charles Darwin. New York: Hill and Wang.

Levine, Alex. 2001. "Individualism, Type Specimens, and the Scrutability of Species Membership." *Biology and Philosophy* 16: 325–38.

Leplin, Jarrett. 1988. "Is Essentialism Unscientific?" *Philosophy of Science* 55: 493–510.

Lewis, David. 1983. "Scorekeeping in a Language Game." In *Philosophical Papers*, vol. I, pp. 233–49. Oxford: Oxford University Press.

1996. "Elusive Knowledge." *Australasian Journal of Philosophy* 74: 549–67.

Li, Chenyang. 1993. "Natural Kinds: Direct Reference, Realism, and the Impossibility of Necessary A Posteriori Truth." *Review of Metaphysics* 47: 261–76.

Liebig, Justus. 1842. *Animal Chemistry*, ed. William Gregory. Cambridge, Mass.: J. Owen. (Facsimile: New York: Johnson Reprint Corporation, 1964.)

Lipman, Timothy O. 1967. "Vitalism and Reductionism in Liebig's Physiological Thought." *Isis* 58: 167–85.

Lipton, Peter. 1999. Review of *Philosophy of Science*, by Alexander Bird. *British Journal for the Philosophy of Science* 50: 163–8.

Locke, John. 1975. *An Essay Concerning Human Understanding*, ed. with a Foreword by Peter Nidditch. Oxford: Oxford University Press.

Logue, James. 1995. "Incommensurability." In *The Oxford Companion to Philosophy*, ed. Ted Honderich, p. 397. Oxford: Oxford University Press.

Lovejoy, Arthur O. 1968. "Buffon and the Problem of Species." In *Forerunners of Darwin, 1745–1859*, ed. B. Glass, O. Temkin, and W. Straus, pp. 84–113. Baltimore: The Johns Hopkins Press.

Lowe, E. J. 1995. "Analytic and Synthetic Statements." In *The Oxford Companion to Philosophy*, ed. T. Honderich, p. 28. Oxford: Oxford University Press.

1997. "Ontological Categories and Natural Kinds." *Philosophical Papers* 26: 29–46.

Lycan, William G. 1994. *Modality and Meaning*. Dordrecht: Kluwer Academic Publishers.

Macbeth, Danielle. 1995. "Names, Natural Kind Terms, and Rigid Designation." *Philosophical Studies* 79: 259–81.

Mackie, J. L. 1976. *Problems from Locke*. New York: Oxford University Press.

Maddy, Penelope. 1990. *Realism in Mathematics*. Oxford: Clarendon Press.

Maor, Eli. 1991. *To Infinity and Beyond*. Princeton, N.J.: Princeton University Press.

Margalit, Avishai. 1979. "Sense and Science." In *Essays in Honour of Jaakko Hintikka*, ed. E. Saarinen, R. Hilpinen, I. Niiniluoto, and M. P. Hintikka, pp. 17–47. Dordrecht: D. Reidel.

Margolis, Eric, and Laurence, Stephen. 2001. "Boghossian on Analyticity." *Analysis* 61, 293–302.

Martin, Graham. 1996. "Birds in Double Trouble." *Nature* 380: 666–7.

Matthen, Mohan. 1998. "Biological Universals and the Nature of Fear." *Journal of Philosophy* 95: 105–32.

Maynard Smith, John. 2001. "Foreword." In *Genes, Categories, and Species*, by Jody Hey, pp. xv–xvii. New York: Oxford University Press.

Mayr, Ernst. 1969. *Principles of Systematic Zoology*. New York: McGraw-Hill.

1970. *Populations, Species, and Evolution: An Abridgment of Animal Species and Evolution*. Cambridge, Mass.: Harvard University Press.

1976. *Evolution and the Diversity of Life: Selected Essays*. Cambridge, Mass.: Harvard University Press.

1982. *The Growth of Biological Thought*. Cambridge, Mass.: Harvard University Press.

1987. "The Ontological Status of Species: Scientific Progress and Philosophical Terminology." *Biology and Philosophy* 2: 145–66.

1995. "Systems of Ordering Data." *Biology and Philosophy* 10: 419–34.

2002. "Comments by Ernst Mayr." *Theory in Biosciences* 121: 99–100.

Mayr, Ernst, and Ashlock, Peter. 1991. *Principles of Systematic Zoology*. Second Edition. New York: McGraw-Hill.

Mayr, E., Linsley, E. G., and Usinger, R. L. 1953. *Methods and Principles of Systematic Zoology*. New York: McGraw-Hill.

McDougall, William. 1934. *Modern Materialism and Emergent Evolution*. Second Edition. London: Methuen.

1938. *The Riddle of Life*. London: Methuen.

McGrew, Tim. 1994. "Scientific Progress, Relativism, and Self-Refutation." *The Electronic Journal of Analytic Philosophy* 2 (2). 19 July 2001 <http://ejap.louisiana.edu/EJAP/1994.may/mcgrew.abs.html>.

Medawar, P. B. and Medawar, J. S. 1983. *Aristotle to Zoos: A Philosophical Dictionary of Biology*. Cambridge, Mass.: Harvard University Press.

Meier, Rudolf, and Willmann, Rainer. 2000. "The Hennigian Species Concept." In *Species Concepts and Phylogenetic Theory*, ed. Quentin Wheeler and Rudolf Meier, pp. 30–43. New York: Columbia University Press.

Mellor, D. H. 1991. "Natural Kinds." In *Matters of Metaphysics*, pp. 123–35. Cambridge: Cambridge University Press.

Miller, Alexander. 1998. *Philosophy of Language*. London: UCL Press.

Miller, Richard B. 1992. "A Purely Causal Solution to One of the Qua Problems." *Australasian Journal of Philosophy* 70: 425–34.

Mishler, B., and Brandon, R. 1987. "Individuality, Pluralism, and the Phylogenetic Species Concept." *Biology and Philosophy* 2: 397–414.

Mishler, B., and Donoghue, M. 1982. "Species Concepts: A Case for Pluralism." *Systematic Zoology* 31: 491–503.

Mondadori, Fabrizio. 1978. "Interpreting Modal Semantics." In *Studies in Formal Semantics*, ed. F. Guenther and C. Rohrer, pp. 13–40. Amsterdam: North-Holland Publishing Co.

Morgan, C. Lloyd. 1923. *Emergent Evolution*. London: Williams and Norgate.

Neale, Stephen. 2001. "No Plagiarism Here: The Originality of Saul Kripke." Review of *The New Theory of Reference*, by Paul Humphreys and James Fetzer, eds. *Times Literary Supplement*, 9 February: 12–13.

Nersessian, Nancy. 1991. "Discussion: The Method to 'Meaning': A Reply to Leplin." *Philosophy of Science* 58: 678–86.

O'Brien, Stephen J. 1987. "The Ancestry of the Giant Panda." *Scientific American* 257 (5): 102–7.

O'Donnell, Kerry, Cigelnik, Elizabeth, and Nirenberg, Helgard I. 1998. "Molecular Systematics and Phylogeography of the Gibberella Fujikuroi Species Complex." *Mycologia* 90: 465–93.

O'Hara, Robert. 1993. "Systematic Generalization, Historical Fate, and the Species Problem." *Systematic Biology* 42: 231–46.

Okasha, S. 2002. "Darwinian Metaphysics: Species and the Question of Essentialism." *Synthese* 131: 191–213.

Olmsted, J. M. D., and Olmsted, E. Harris. 1952. *Claude Bernard and the Experimental Method in Medicine*. New York: Henry Schuman.

"Onion"a. 2001. *The Columbia Encyclopedia*, Sixth Edition. New York: Columbia University Press. 21 June 2001 <http://www.bartleby.com/65>.

"Onion"b. *Encyclopædia Britannica Online*. 21 June 2001 <http://search.eb.com/eb/article?eu=58571>.

Osborn, Henry Fairfield. 1905. "Tyrannosaurus and Other Cretaceous Carnivorous Dinosaurs." *Bulletin of the American Museum of Natural History* 21: 259–65.

Ottobre, Ron. 2001. "Fresh Asparagus: A Gift from Spring." *Baltimore Sun*, 11 April. LEXIS-NEXIS. 15 June 2001 <http://web.lexis-nexis.com/universe>.

Papineau, David. 1996. "Theory-Dependent Terms." *Philosophy of Science* 63: 1–20.

Paterson, H. 1985. "The Recognition Concept of Species." In *Species and Speciation*, ed. E. Vrba, pp. 21–9. Pretoria: Transvaal Museum.

Patterson, Colin. 1993. "Scientific Correspondence." *Nature* 366: 518.

Plantinga, Alvin. 1993. *Warrant and Proper Function*. Oxford: Oxford University Press.

Platts, Mark. 1997. *Ways of Meaning: An Introduction to a Philosophy of Language*. Second Edition. Cambridge, Mass.: MIT Press.

Philippe, Hervé. 1997. "Rodent Monophyly: Pitfalls of Molecular Phylogenies." *Journal of Molecular Evolution* 45: 712–15.

Poncinie, Lawrence. 1985. "Meaning Change for Natural Kind Terms." *Noûs* 19: 415–27.

Pough, F. Harvey, Heiser, John B., and McFarland, William N. 1996. *Vertebrate Life*. Fourth Edition. Upper Saddle River, N.J.: Prentice Hall.

Putnam, Hilary. 1975a. *Mind, Language and Reality: Philosophical Papers*, vol. 2. Cambridge: Cambridge University Press.

1975b. "The Analytic and the Synthetic." In *Mind, Language and Reality: Philosophical Papers*, vol. 2, pp. 33–69. Cambridge: Cambridge University Press.

1975c. "Is Semantics Possible?" In *Mind, Language and Reality: Philosophical Papers*, vol. 2, pp. 139–52. Cambridge: Cambridge University Press.

1975d. "Explanation and Reference." In *Mind, Language and Reality: Philosophical Papers*, vol. 2, pp. 196–214. Cambridge: Cambridge University Press.

1975e. "The Meaning of 'Meaning'." In *Mind, Language and Reality: Philosophical Papers*, vol. 2, pp. 215–71. Cambridge: Cambridge University Press.

1975f. "Language and Reality." In *Mind, Language and Reality: Philosophical Papers*, vol. 2, pp. 272–90. Cambridge: Cambridge University Press.

1983a. "Possibility and Necessity." In *Realism and Reason: Philosophical Papers*, vol. 3, pp. 46–68. Cambridge: Cambridge University Press.

1983b. "'Two Dogmas' Revisited." In *Realism and Reason: Philosophical Papers*, vol. 3, pp. 87–97. Cambridge: Cambridge University Press.

1992. "Is It Necessary That Water Is $H_2O$?" In *The Philosophy of A. J. Ayer*, ed. L. E. Hahn. La Salle, Ill.: Open Court.

Quaker Oats Company. 1997. "Did U Know?" Chicago, Ill.: Quaker Oats Co.

Quine, W. V. 1961. "Two Dogmas of Empiricism." In *From a Logical Point of View*, pp. 20–46. Cambridge, Mass.: Harvard University Press.

1976a. "Truth by Convention." In *Ways of Paradox and Other Essays*, pp. 77–106. Cambridge, Mass.: Harvard University Press.

1976b. "Carnap and Logical Truth." In *Ways of Paradox and Other Essays*, pp. 107–32. Cambridge, Mass.: Harvard University Press.

Reid, Jasper. 2002. "Natural Kind Essentialism." *Australasian Journal of Philosophy* 80: 62–74.

Reimer, Marga. 1997. "Could There Have Been Unicorns?" *International Journal of Philosophical Studies* 5: 35–51.

Rey, Georges. 1993. "The Unavailability of What We Mean: A Reply to Quine, Fodor, and Lepore." *Grazer Philosophische Studien* 46: 61–101.

Ridley, Mark. 1986. *Evolution and Classification: The Reformation of Cladism*. London: Longman.

1989. "The Cladistic Solution to the Species Problem." *Biology and Philosophy* 4: 1–16.

Ritvo, Harriet. 1993. "Zoological Taxonomy and Real Life." In *Realism and Representation*, ed. G. Levine, pp. 235–54. Madison: University of Wisconsin Press.

Rockwell, F. F., Grayson, Esther C., and de Graaff, Jan. 1961. *The Complete Book of Lilies*. Garden City, N.Y.: Doubleday.

Rorty, Richard. 1999. Review of *The Social Construction of What?*, by Ian Hacking. *The Atlantic Monthly*, 284 (5): 120–2.

Rosen, Donn Eric. 1974. "Cladism or Gradism?: A Reply to Ernst Mayr." *Systematic Zoology* 23: 446–51.

Rosenberg, Alexander. 1987. "Why Does the Nature of Species Matter?" *Biology and Philosophy* 2: 192–7.

Rosenfeld, Stuart, and Bhushan, Nalini. 2000. "Chemical Synthesis: Complexity, Similarity, Natural Kinds, and the Evolution of a 'Logic'." In *Of Minds and Molecules: New Philosophical Perspectives on Chemistry*, ed. Bhushan and Rosenfeld, pp. 187–207. Oxford: Oxford University Press.

Rosselló-Mora, Ramon, and Amann, Rudolf. 2001. "The Species Concept for Prokaryotes." *FEMS Microbiology Reviews* 25: 39–67.

Rudler, Frederick William. 1911. "Jade." In *Encyclopaedia Britannica*, Eleventh Edition. vol. 15, pp. 122–4.

Ruse, Michael. 1987. "Biological Species: Natural Kinds, Individuals, or What?" *British Journal for the Philosophy of Science* 38: 225–42.

1999. *Mystery of Mysteries: Is Evolution a Social Construction?* Cambridge, Mass.: Harvard University Press.

Russell, E. S. 1911. "Vitalism." *Rivista di scienza, "Scientia"* 9: 329–45.

1930. *The Interpretation of Development and Heredity: A Study in Biological Method.* Oxford: Clarendon Press.

Ryder, Oliver. 1987. "The Giant Panda Is a Bear." *Zoonooz* 60 (8): 16–17.

Sainsbury, Mark. 1991. *Logical Forms.* Oxford: Basil Blackwell.

Sakikawa, Noriyuki. 1968. *Jade.* Tokyo: Japan Publications.

Salmon, Merrilee, et al. 1992. *Introduction to the Philosophy of Science.* Englewood Cliffs, N.J.: Prentice Hall.

Salmon, Nathan. 1980. *Reference and Essence.* Princeton, N.J.: Princeton University Press.

Sankey, Howard. 1994. *The Incommensurability Thesis.* Aldershot: Avebury.

Scheffler, Israel. 1967. *Science and Subjectivity.* Indianapolis: Bobbs-Merrill.

Schloegel, Judy Johns. 1999. "From Anomaly to Unification: Tracy Sonneborn and the Species Problem in Protozoa, 1954–1957." *Journal of the History of Biology* 32: 93–132.

Schultz, Stanley. 1998. "A Century of (Epithelial) Transport Physiology: From Vitalism to Molecular Cloning." *American Journal of Physiology* 274: C13–C23.

Schwartz, Steven P. 1977. "Introduction." In *Naming, Necessity, and Natural Kinds*, pp. 13–41. Ithaca, N.Y.: Cornell University Press.

1980. "Formal Semantics and Natural Kind Terms." *Philosophical Studies* 38: 189–98.

"Second General Discussion Session." 1974. *Synthese* 27: 509–21.

Segal, Gabriel. 2000. *A Slim Book About Narrow Content.* Cambridge, Mass.: MIT Press.

Shapere, Dudley. 1981. "Meaning and Scientific Change." In *Scientific Revolutions*, ed. Ian Hacking, pp. 28–59. Oxford: Oxford University Press.

1998. "Incommensurability." In *Routledge Encyclopedia of Philosophy*, vol. 4, ed. Edward Craig, pp. 732–6. London: Routledge.

Sidelle, Alan. 1989. *Necessity, Essence, and Individuation: A Defense of Conventionalism.* Ithaca, N.Y.: Cornell University Press.

Simpson, G. G. 1961. *Principles of Animal Taxonomy.* New York: Columbia University Press.

Smith, Hobart M. 1990. "The Universal Species Concept." *Herpetologica* 46: 122–4.

Sneath, Peter, and Sokal, Robert. 1973. *Numerical Taxonomy: The Principles and Practice of Numerical Classification*. San Francisco: W. H. Freeman and Co.

Sober, Elliott. 1984. *The Nature of Selection: Evolutionary Theory in Philosophical Focus*. Cambridge, Mass.: MIT Press.

Sokal, Robert, and Sneath, Peter. 1963. *Principles of Numerical Taxonomy*. San Francisco: W. H. Freeman and Co.

Spears, Tom. 1993. "Modern Dinosaurs End Up on Dinner Plate." *Ottawa Citizen*, 29 December. LEXIS-NEXIS. 26 April 2001 <http://web.lexis-nexis.com/universe>.

Speer, Brian R. 1998. "Liliales: The True Lilies." University of California, Berkeley Museum of Paleontology. 15 June 2001 <http://www.ucmp.berkeley.edu/monocots/liliflorae/liliales.html>.

Spencer, Leonard J. 1936. *A Key to Precious Stones*. London: Blackie & Son Limited.

Stamos, David. 1998. "Buffon, Darwin, and the Non-Individuality of Species: A Reply to Jean Gayon." *Biology and Philosophy* 13: 443–70.

Sterelny, Kim. 1983. "Natural Kind Terms." *Pacific Philosophical Quarterly* 64: 110–25.
——— 1994. "The Nature of Species." *Philosophical Books* 35: 9–20.

Stevens, P. F. 1984. "Metaphors and Typology in the Development of Botanical Systematics 1690–1960, or the Art of Putting New Wine in Old Bottles." *Taxon* 33: 169–211.

Steward, Helen. 1990. "Identity Statements and the Necessary A Posteriori." *Journal of Philosophy* 87: 385–98.

Strickberger, Monroe W. 1996. *Evolution*. Second Edition. Sudbury, Mass.: Jones and Bartlett.

Swoyer, Chris. 2001. "Properties." *The Stanford Encyclopedia of Philosophy* (Spring 2001), ed. Edward N. Zalta. August 31, 2001 <http://plato.stanford.edu/archives/spr2001/entries/properties/>.

Tangley, Laura. 1997. "At Last, the Missing Link." *U.S. News & World Report*, June, 14.

Teich, Mikuláš, and Needham, Dorothy. 1992. *A Documentary History of Biochemistry, 1770–1940*. Leicester: Leicester University Press.

Teller, Paul. 1977. "Indicative Introduction." *Philosophical Studies* 31: 173–95.

Templeton, Alan. 1992. "The Meaning of Species and Speciation: A Genetic Perspective." In *The Units of Evolution*, ed. M. Ereshefsky, pp. 159–83. Cambridge, Mass.: MIT Press.

Tennant, Smithson. 1797. "On the Nature of the Diamond." *Philosophical Transactions of the Royal Society of London* 87 (97): 123–7.

Tooley, M. 1977. "The Nature of Laws." *Canadian Journal of Philosophy* 7: 667–98.

Unger, Peter. 1975. *Ignorance: A Case for Scepticism*. Oxford: Oxford University Press.

Van Brakel, J. 1992. "Natural Kinds and Manifest Forms of Life." *Dialectica* 46: 243–61.

Van den Hoek, C., Mann, D. G., and Jahns, H. M. 1995. *Algae*. Cambridge: Cambridge University Press.

Van Valen, L. 1976. "Ecological Species, Multispecies, and Oaks." *Taxon* 25: 233–9.
——— 1989. "Metascience." *Evolutionary Theory* 9: 99–103.

Waismann, Friedrich. 1945. "Verifiability." *Proceedings of the Aristotelian Society* 19: 119–50.

Ward, Fred, and Ward, Charlotte. 1996. *Jade*. Bethesda, Md.: Gem Book Publishers.

Watson, Hewett C. 1845. "On the Theory of 'Progressive Development,' Applied in Explanation of the Origin and Transmutation of Species." *Phytologist* 2: 108–13, 140–7.

Wheeler, L. R. 1939. *Vitalism: Its History and Validity*. London: Witherby.

Whitlock, Herbert, and Ehrmann, Martin. 1949. *The Story of Jade*. New York: Sheridan House.

Wiley, E. O. 1978. "The Evolutionary Species Concept Reconsidered." *Systematic Zoology* 27: 17–26.

1981. *Phylogenetics*. New York: Wiley.

Wilkerson, T. E. 1993. "Species, Essences and the Names of Natural Kinds." *The Philosophical Quarterly* 43: 1–19.

Williams, James. 1995. "Pioneer in a Random World." *Times* (London), 12 May, Educational Supplement, p. xvi.

Wills, Geoffrey. 1964. *Jade*. London: Arco Publications.

Wilson, Jack. 1999. *Biological Individuality: The Identity and Persistence of Living Entities*. Cambridge: Cambridge University Press.

Wolfram, Sybil. 1989. *Philosophical Logic: An Introduction*. London: Routledge.

Wollaston, Thomas Vernon. 1973. Review of the *Origin of Species*. Reprinted in *Darwin and His Critics: The Reception of Darwin's Theory of Evolution by the Scientific Community*, ed. David Hull, pp. 127–40. Cambridge, Mass.: Harvard University Press.

# Index

Scheffler, Israel, 191n6
Schliesser, Eric, xii
Schloegel, Judy Johns, 74
Schultz, Stanley, 142, 195n29
Schwartz, Steven P., 39, 48, 183nn6,8
scientific revolutions, 121, 126, 128, 134, 142, 194n23
Segal, Gabriel, 175n4
Shapere, Dudley, 117, 191n5, 195n31, 196n6
Sidelle, Alan, 165, 183n7
Simpson, G. G., 67, 77, 86–7, 186n15
skepticism
    and fallibilism, possibility of revision (*see* unrevisability)
    of incommensurability theorists, 112, 128–9, 135, 137 (*see also* progress, scientific)
Smith, Hobart, 73
Smith, Peter, 193n19, 196nn7,8
Sneath, Peter, 186n17
Sober, Elliot, 162, 197n14
social construction, science as, 140–1
Sokal, Robert, 186n17
special creationism. *See* Darwinian revolution
species, biological
    baptism of 'species', 121
    discovering essences of, 70–6, 85–7
    distinguished from higher taxa, 175n1
    and essentialism, 11–13, 52–62
    as historically connected, 10–12
    as individuals, 9–17
    as kinds, 9–17
    and special creationism, evolutionism, 121–35, 147–8, 155–6
    and vagueness, 156–9
    *See also* taxa
species of minerals, 102
Stamos, David, 176n3, 178n10
statements, 161
Sterelny, Kim, 94, 175n6, 176n3, 191n7, 198n17
Stevens, P. F., 186n18
Steward, Helen, 109
Stjernfelt, Frederik, 139
straight lines. *See* triangles
*Structure of Scientific Revolutions*, 191n11
Swoyer, Chris, 177n8
synonymy, 161–8, 197nn13,15
synthetic statements. *See* analyticity
systematics. *See* taxonomy

taxa, biological
    higher taxa distinguished from species, 175n1
    laws about, 13–14

naturalness of, 17–27 *passim*, 72–3, 81, 186n19
    ranking, 175n1, 180n21, 186n19
    *See also* higher taxa; species
taxonomy
    and cartography, 80–1
    and competing schools, 20–2, 70–83
    *See also* cladism; pattern cladism; pheneticism
Teller, Paul, 94
Templeton, Alan, 86
Tennant, Smithson, 101
theoretical identity statements. *See* identity statements: theoretical
theory change
    absence of (*see* language change: without theory change)
    distinguished from language change (*see* meaning change–theory change distinction)
    with language change, 90–1, 111, 146–7, 150–9, 172
    without language change, 129–31, 135–6, 147 (*see also* discovery)
theorymeaning change, 151, 155
*Thunder-Lizardosaurus*, 166–7
tigers, 51–62 *passim*, 71–2
Tooley, M., 14
topaz, 102
translation
    and incommensurability, 143–6
    indeterminacy of, 151
triangles, 151–4, 159
truth
    by convention, 163
    by definition, 122, 127, 155, 158
    degrees of, on precisifications, etc., 131, 157, 193n20, 198n16, 199n25
    logical, 160–1, 197n13
    and scientific progress, 128–9, 194n23
Twin Earth. *See* XYZ
type specimens, 5–6, 184n14
'*Tyrannosaurus rex*'
    coining of, 5–7
    *See also* birds

Unger, Peter, 23
unrevisability, 161–3, 168, 171–2

vagueness
    accounts of, 131, 157, 193n20
    and conflicting conditions for reference, 123–5, 156–8, 196nn6,9, 197n10
    on different theories of reference, 116, 118–20
    and inevitability of scientific conclusions, 140